AF583357

Environmental Defenders

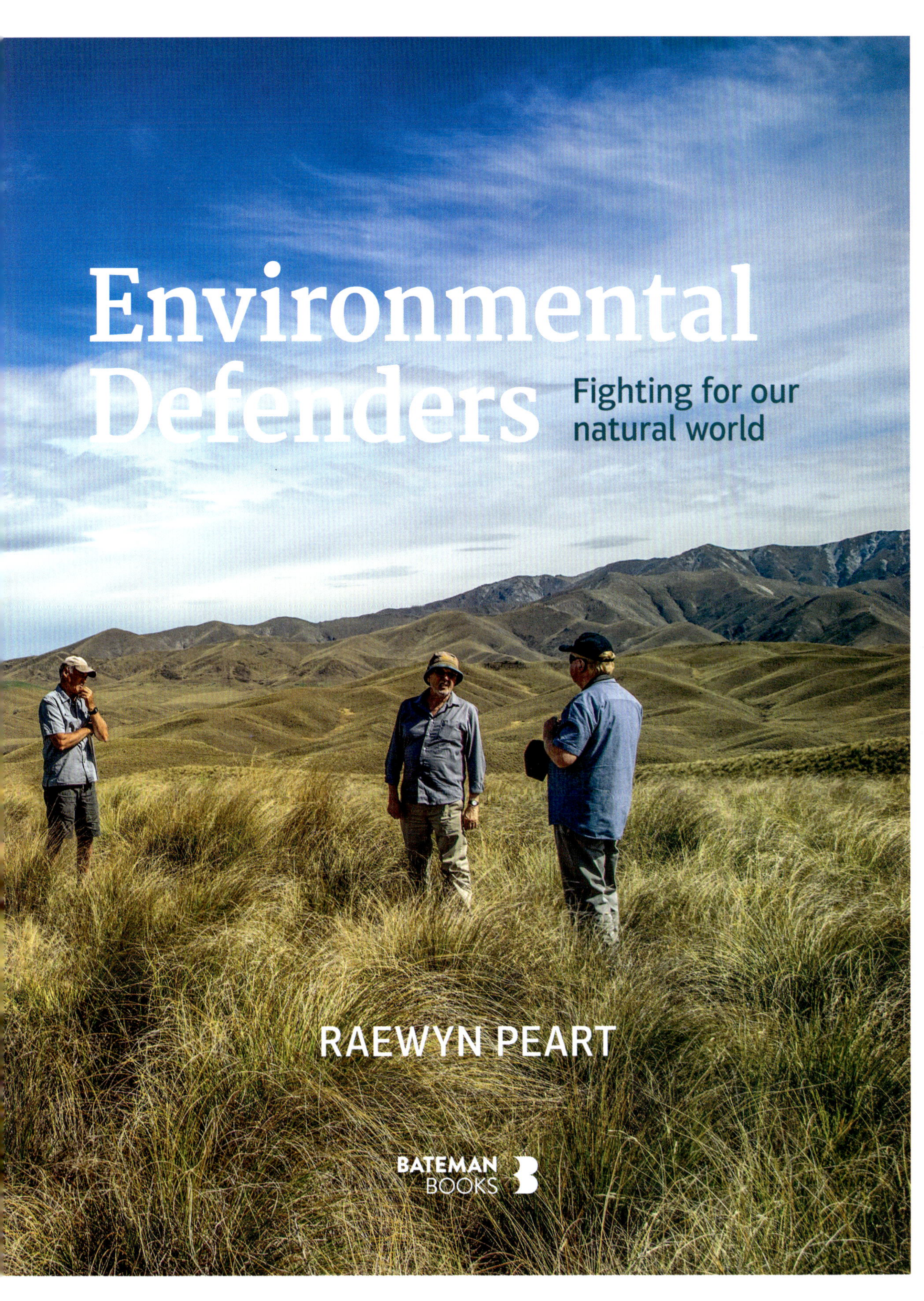
Environmental Defenders
Fighting for our natural world
RAEWYN PEART
BATEMAN BOOKS

Publisher's note: Where there is a Māori place name, both the Māori name and the European name are given on first use in each chapter. After that, the most commonly used name is applied.

Published in 2024 by David Bateman Ltd,
Unit 2/5 Workspace Drive, Hobsonville,
Auckland 0618, New Zealand
www.batemanbooks.co.nz

ISBN: 978-1-77689-100-9

A catalogue record for this book is available from the National Library of New Zealand.

Book design: Alice Bell
Printed in China by Everbest Printing Investment Limited

Previous pages: EDS field trip to the Otago high country. RAEWYN PEART
Front cover: Tussock lands, Te Manahuna Mackenzie Basin. ANDRIS APSE
Back cover: Waikawau Bay, Te Tara-o-te-Ika-a-Māui Coromandel Peninsula. RAEWYN PEART
Spine: Pakiri Beach, north of Tāmaki Makaurau Auckland. RAEWYN PEART
Contents: Te Henga Bethells Beach, Waitākere Ranges Regional Park. RAEWYN PEART
Part 1 (pages 14–15): Mata-au Clutha River with morning mist. ROB SUISTED/NATURESPIC
Part 2 (pages 168–69): Te Henga Bethells Beach, Waitākere Ranges Regional Park. RAEWYN PEART

This book is dedicated to Gary Taylor, who has been the main impetus behind the Environmental Defence Society since the late 1970s, and who is a valued colleague, mentor and friend.

Contents

Foreword

Even for someone who lived through the events described, it is a shock now to be reminded how undeveloped environmental protection was when the Environmental Defence Society was founded in 1971 to promote it through public education, law reform and legal action. It was a time when development was widely regarded as a public good and was driven by major government departments. There was little urgency about the preservation of unmodified ecological systems and landscapes. Bad practices in sewage discharges, rubbish disposal, and agricultural and industrial chemical use often went unchallenged. Environmental legislation was undeveloped. Access to the courts for public interest litigants was hampered by restrictions on standing and the late development of modern administrative law.

Fifty years later, Raewyn Peart's history of the society is a highly engaging account of environmental causes and the people who came together to fight for them. The experiences described contain important lessons for environmental protection today and the challenges it will have to meet. It is a story of imagination and verve, illustrated by beautiful photographs of the landscapes and waters of New Zealand. It charts a cultural shift in society and law in relation to environmental protection, an astonishing amount of which was led or prodded by EDS.

The cultural shift that has taken place in the last 50 years is underscored by the different conditions and challenges faced by EDS when it was revived in 1999 after a gap of more than a decade. These changed circumstances make the history of EDS a history in two parts. Although there was reason for optimism in 1988, the hiatus served to demonstrate the need for an independent advocate for the environment willing to tackle hard issues like incremental degradation and landscape protection. Indeed it is tempting to speculate whether some of the missteps in interpretation and application of the Resource Management Act would have taken root if EDS had been active when the Act came into effect.

The new climate in 1999 included a swing in public support for environmental protection. There were, however, new challenges requiring some shift in focus and method. New sources of funding, private as well as public, have made possible more professional administration of conferences and events and more professional policy direction including producing and commissioning high-quality research and publications to support the extremely diverse projects undertaken. As in the original conception in 1971, litigation is only one in a range of activities undertaken by EDS. Law reform remains an important object, as does public education and practical assistance, and a willingness to cooperate with others, including business, developers, farming and government, to secure practical solutions of benefit to the environment.

EDS brings a number of advantages to its defence of the environment which have

Kenepuru Sound, Marlborough. RAEWYN PEART

remained constant throughout and are illustrated in the book. I mention four only.

Because EDS is concerned with the environment as a whole, it has had experience over an extraordinarily wide range of issues and with many different communities. That gives it a unique overview of the system which allows it to pick the issues that matter, including those concerned with the architecture of protection.

While many of its causes have concerned matters of national significance, EDS has always paid close attention to local issues and the concerns of local communities. It has been willing to help where it has expertise and where small groups would otherwise be swamped by the scale of the challenges they face. These experiences keep it grounded.

EDS was designed to draw on scientists and other professional disciplines as well as lawyers (although in my experience the scientists were pretty good lawyers too). The scientists, engineers, economists, social scientists and historians involved in EDS (as directors, policy analysts, researchers and authors of think-pieces) have been of outstanding quality and have ensured intellectual rigour and a focus on evidence. Similarly, and starting with Sir David Williams, the lawyers associated with EDS have been a roll call of New Zealand legal talent. The effectiveness of EDS reflects the quality of its work and the standing of those who have contributed to it.

Finally, I mention the manner in which EDS operates. It is cheerful, purposeful, practical and respectful of others. That is the tone set since his arrival in 1978 by Gary Taylor. It is why those who have worked with EDS and whose stories are told in this book count the experience among the highlights of their lives, as I do.

The Right Honourable Dame Sian Elias GNZM KC

Preface

The Environmental Defence Society (EDS) was established over 50 years ago, in April 1971. In its first phase, it was operational until 1988, when it went into recess. EDS was re-established in 1999 and is still active today. This book has been written to reflect on and celebrate the achievements of the society over those years.

The book is an insider account written by long-time EDS director and policy director Raewyn Peart. It is primarily based on EDS records and newsletters, reported court cases and interviews with many current and former EDS directors and staff. It is therefore not an independent history of the society. Nor is it an account of the considerable contributions of other environmental organisations within Aotearoa New Zealand. Instead, the book provides a view from within EDS of what it has been endeavouring to achieve over the years, and how it has gone about it. In that sense it can be regarded as an 'autobiography' of the society.

The focus of EDS has been on using the law to achieve improved environmental outcomes. It is hoped this book, which chronicles EDS's experiences, will provide some insight into how the law has been mobilised for the benefit of nature in the past and how it might be used in this way into the future.

Raewyn Peart

Takapō Tekapo River, Te Manahuna Mackenzie Basin.
RAEWYN PEART

Acknowledgements

I would like to thank Sir David Williams, who provided financial assistance for this book, and Rob Suisted, Craig Potton, GNS Science, David Hartman, Shellie Evans, Project Kahurangi, Alexander Turnbull Library and the Auckland Libraries, who provided images for this book at no or very reduced cost. Mary Binney kindly provided consent to reproduce Don Binney's artwork which he created to support the Karikari campaign. Bell Gully provided a pro bono legal review of the manuscript undertaken by Kristin Wilson and Sarah Broughan, which was greatly appreciated.

I would also like to thank the many people who agreed to be interviewed for the book. They include Barry Barton, Dick Bellamy, Malcolm Bowman, Richard Brabant, Andrew Brown, Stephen Brown, John Burns, Ian Cowper, Catherine Delahunty, Nicola de Wit, Marie Doole, Fiona Driver, Elizabeth Edmonds, Rob Enright, Sarah Foreman, Natasha Garvan, Bryce Johnson, Shona Kenderdine, David Kirkpatrick, Deidre Koolen-Bourke, Garry Law, Graeme Lawrence, John Leader, Di Lucas, Bob Mann, David McGregor, Stephen Mills, Laurie Newhook, Mike O'Sullivan, David Parker, Tanya Peart, Evan Penny, Tony Randerson, Peter Reaburn, Guy Salmon, Peter Salmon, Rowan Sapsford, Shay Schlaepfer, Greg Severinsen, Nick Smith, Rosalie Snoyick, Kate Storer, Gary Taylor, Denis Tegg, Mark Tugendhaft, Susan Walker, David Williams, Neale Wills, Cordelia Woodhouse and Madeleine Wright.

All errors and omissions remain the responsibility of the author. I have endeavoured to check the accuracy of the content and to obtain permissions to reproduce all the images used, but please advise any issues and I will endeavour to remedy these in any subsequent editions.

Old pōhutukawa at Wainuiototo New Chums Beach, Te Tara-o-te-Ika-a-Māui Coromandel Peninsula.
RAEWYN PEART

PART I

Beginnings: 1971 to 1988

1. Environmental defender needed

IN THE EARLY 1970s, New Zealand society was heavily regulated, conservative judges presided over the courts and there was virtually no public involvement in government decision-making. As Andrew Brown recalls, 'there was a cosy government sector, with no oversight, and a quiescent public. At the same time, there was a burst of energy from younger lawyers who were coming out of law school in greater numbers, and who were excited by prospects in new fields. That is where the idea of EDS [the Environmental Defence Society] fell on fertile ground.'

THE ENVIRONMENTAL DEFENCE Society (EDS) officially came into existence on 22 April 1971, when the seal of the Assistant Registrar of Incorporated Societies was affixed to an application lodged by 30-year-old lawyer David A. R. Williams. Williams had recently returned from post-graduate legal studies in the United States and had become a partner in Tāmaki Makaurau Auckland corporate law firm Russell McVeagh McKenzie Bartleet and Co (known as Russell McVeagh).

The same year saw the completion of the Manapōuri hydro power station in Te Rua-o-Te-Moko Fiordland and the commissioning of the Tiwai Point aluminium smelter near the entrance to Bluff Harbour. Controversy over proposals to raise the level of Lake Manapōuri, in order to provide electricity for the smelter, had galvanised many in the small nation of less than 3 million to oppose the plans and kick-started New Zealand's nascent environmental movement.

But EDS's beginnings had a different impetus.

David Williams, who was born in Auckland during the Second World War, had shown little academic promise during his school years. He initially failed the university entrance exam, before eventually passing and being accepted into the University of Auckland's Law School. After graduating with an LLB, Williams was one of only a handful of New Zealand law graduates at that time who sought to pursue post-graduate study overseas, and one of even fewer who headed to an American university. New Zealand's intellectual ties were very much with Britain at that time. As Williams recalled,

Previous pages: Overlooking David Wiliams's property at Pakiri Beach, north of Auckland. RAEWYN PEART

In those days it was a very unusual thing to study in the United States. The general view of the legal profession in Auckland was that you tried to get into a good local law firm as soon as possible. It was a waste of time to study overseas, but if you did, there were only two places to go — Oxford or Cambridge.

But the Dean of Auckland Law School, Professor Jack Northey, was both knowledgeable and enthusiastic about American universities, so I became interested in going to an American law school. I needed a scholarship to go, as I had no money, but my grades were not particularly good. Nevertheless, I applied to six law schools and five turned me down. Then, to my amazement, I was offered a scholarship to attend Harvard Law School.

When I arrived at Harvard I was interviewed by the Foreign Student Advisor who said 'would you like to know why you are here Mr Williams? Your law school grades were not spectacular but you have been President of the Auckland University Law School Students Association. Moreover, Harvard aspires to be the greatest law school in the world and likes to have international coverage. This year, no one apart from you applied from New Zealand or Australia.' It was deflating, but I didn't care. I was there.

At Harvard there was a great choice of subjects. One in particular caught my eye – environmental law. While I was there, I heard about the Environmental Defense Fund (EDF) and its DDT litigation. Charles Wurster, a science professor from New York who had been one of the founders of EDF, gave a lecture at Harvard Law School. He explained that DDT caused a lot of negative human-health impacts as well as the poisoning of birds. I got curious and read

> up about the EDF approach. Later I wrote to Professor Wurster explaining that I was about to leave New York to head home, and he replied 'why not come down to Long Island' where he was teaching.
>
> So I went there and had a long chat with him. We discussed the EDF concept of combining legal and scientific knowledge and he encouraged me to pursue the idea when I returned home. He was very supportive and encouraging.

Professor Emeritus Charles F. Wurster was one of the founders of EDF and from 1999 until his death in July 2023 he was a director of EDS. He earned a doctorate in chemistry from Stanford University in 1957 and for 30 years was Professor of Environmental Sciences at the State University of New York at Stony Brook. Being an ardent ornithologist, Wurster became concerned about the environmental effects of the insecticides DDT, aldrin and dieldrin during the 1960s. Under the umbrella of EDF, Wurster and his colleagues challenged their use through the courts and by 1974 all three were banned in the United States.

Williams returned to New Zealand in 1966. While he set about establishing his legal career, the idea of setting up an EDF-type organisation in New Zealand remained at the back of his mind.

—ooo—

DURING THE LATE 1960s, there was an increasing awareness in New Zealand of the impacts of post-war economic growth on the country's natural environment. In 1968, the Holyoake-led National government established the Physical Environment Committee which was tasked with considering the 'kind of physical environment to which the community might aspire' and problems associated with attaining it. Matters within the committee's ambit included water and air pollution and 'the preservation of natural scenic and recreational endowments'. The committee called for a conference on the physical environment and this was convened by the Ministry of Works in May 1970. David Williams attended the event.

At the conference, Chief Ombudsman Sir Guy Powles, who had earlier trained as a lawyer, gave a powerful presentation. He highlighted the multitude of public bodies that had been created to 'ensure control of the environment for the public benefit' but which seemed reluctant to enforce compliance with the law. He queried how ordinary citizens might 'compel or induce or encourage public bodies adequately to look after our environment' through taking legal action, concluding that, 'this right, as our law stands at present, is of a limited nature, and it would be a wise and bold lawyer who would venture an opinion on its extent and effectiveness, or even on how to go about using it. A citizen does have some right to bring before the Courts a public body which has done something wrong, but if it has done nothing he has no remedy.'

The presentation struck a strong chord with David Williams, who already had a deep interest in environmental justice. He later invited Sir Guy to become the patron of EDS.

After his return to New Zealand from Harvard, Williams had seen numerous examples of shoddy environmental practices. Raw sewage was being illegally pumped into the Waikato River, rubbish was being dumped onto Auckland foreshores and toxic chemicals, such as the herbicide 2,4,5-T, were being

DAVID HARTMAN

Sir David Williams KNZM KC* was born in Auckland in 1941. In 1965 he graduated from Auckland University with an LLB and gained an LLM from Harvard University the following year. He became a litigation partner at Russell McVeagh in 1969. In 1974, Williams became inaugural lecturer of the first environmental law course offered for the Bachelor of Laws degree at the University of Auckland. Williams went on to write the first New Zealand textbook on environmental law in 1980. That work, now entitled *Environmental and Resource Management Law* (Lexis Nexis), is in its 7th edition, with Derek Nolan KC now editor.

Williams left Russell McVeagh in 1984 to become a barrister. In 1987 he was appointed a Queen's Counsel and he became a High Court Judge in 1991. He also sat on the Cook Islands High Court and Court of Appeal.

In 1994, Williams left the New Zealand bench and continued to work as a barrister. In subsequent years, he established a highly successful career as an international arbitrator, sitting on tribunals around the world that decided billion-dollar disputes. In 2017, he was awarded a knighthood in recognition of his outstanding contribution to the development of arbitration law and practice in New Zealand and internationally.

Williams established EDS in 1971 and was a key figure behind the development of environmental law in New Zealand. He gave the society much of its initial impetus and was on the EDS Board until 1974. Intellectual property law expert Andrew Brown KC, who in 1971 was one of EDS's first student directors, described him as 'a trailblazer and also a wonderful mentor of young talent. He helped mentor me and so many other young lawyers. He was driven and he worked really hard. I would often go to work in the weekends with him which was not the done thing. I vividly remember him in his vertically striped white and blue pants! He cut a good figure and he was very motivating. You were enthused by his passion for the subject. He single-handedly motivated a whole generation really.'

Williams is a keen surfer and cyclist and has a house at Pakiri Beach on Auckland's east coast. He maintains his interest in the environment and recently became involved in legal proceedings to stop sand mining off Pakiri Beach through the activist local environmental group Friends of Pakiri Beach Incorporated. EDS is also involved in the litigation opposing the sand mining.

* With the death of Her Majesty Queen Elizabeth II all lawyers with the title Queen's Counsel (QC) are now King's Counsel (KC). Where they are referred to in a historic context, QC will be retained.

liberally sprayed around the countryside. He also saw shortcomings in the array of environmental organisations then operating within the country, with long-established pro bono groups such as the Royal Forest and Bird Protection Society (Forest and Bird), Civic Trust and acclimatisation societies being stymied by a lack of legal skills. He had observed the effectiveness of EDF in the United States and it was clear to him that there needed to be a similar organisation in New Zealand.

Williams explained shortly after EDS's establishment.

> With these ideas in mind we formed the Environmental Defence Society Incorporated in April 1971. It is a coalition of lawyers, scientists, students, and other citizens dedicated to the protection of environmental quality through public education and legal action. It engages in research, presents submissions to appropriate authorities including committees at Parliament, institutes a small number of carefully chosen actions, publishes studies in environmental law, and works to increase public awareness of environmental problems. In essence it is designed to provide countervailing pressure to overcome the inaction of government and its agencies and the intransigence of industry.

EDS remained true to this initial vision for more than half a century, and it continues to do so.

—ooo—

FROM THE VERY BEGINNING, EDS brought together a mix of lawyers, scientists and other professionals to fight for environmental causes. Sir Guy Powles, New Zealand's first ombudsman, agreed to become patron. The organisation quickly developed a close relationship with the University of Auckland, inviting Auckland law school students to join the EDS board, and drawing heavily on the university's scientific expertise.

One of EDS's first actions was aimed at cleaning up the recently commissioned chimney at Auckland Hospital, which was belching out copious amounts of smoky fumes. EDS members lodged numerous complaints with smoke inspectors at Auckland City Council and these eventually resulted in the closure of the chimney while proper smoke-control devices were fitted.

It was not until the passing of the Clean Air Act in 1972 that New Zealand had comprehensive legislation dealing with air pollution. Coordinated by EDS student director Andrew Brown, a group of law students drafted submissions on the Bill on behalf of EDS, and these resulted in several important changes. The changes meant that interested parties were able to make representations to the Clean Air Council, members of the public were guaranteed access to air pollution licence records, and the Clean Air Council was required to file an annual report in Parliament.

The establishment of EDS was given further impetus when Charles Wurster visited New Zealand on a lecture tour in early 1972. He met with EDS directors several times, and when they expressed concern that the society lacked legal standing to sue the Crown, Wurster replied 'do it anyway, just keep suing your government'. EDS directors kept these words in mind as they looked for legal avenues to hold government to account.

—ooo—

Auckland Hospital under construction showing the new chimney. Numerous complaints by EDS about smoky fumes forced the chimney's closure while smoke-control devices were fitted.
NATIONAL PUBLICITY STUDIOS, AUCKLAND LIBRARIES

THE FIRST COURT CASE initiated by EDS concerned the Waikato River, the country's largest river and, at that time, the most polluted. The immense waterway, once rich in aquatic life and a taonga to Tainui and Ngāti Tūwharetoa iwi, was being treated as little more than a waste dump. Accompanied by University of Auckland freshwater ecology doctoral student Wayne Donovan, David Williams headed south to investigate whether things were as bad as reported in the media. Their discussion with the Hamilton-based Medical Officer of Health confirmed that, indeed, pollution of the river was at alarming levels. So Williams looked for a potential court case to highlight the issue.

By that time, the discharge of raw sewage into the river by Huntly Borough Council had been authorised by a series of temporary permits for over 16 years, with no sign of a treatment plant being built. The permits had been liberally granted by the Pollution Advisory Council, a ministerially appointed quango, under the Waters Pollution Act 1953.

There was some doubt as to the status of the last temporary permit held by the council, which appeared to have expired in February 1972. Shortly thereafter, on 1 April 1972, the Water and Soil Conservation Amendment Act (No 2) 1971 came into force. The council was advised by the new regional water board, which processed water discharge permits under the new statutory provisions, that renewal was unnecessary on the basis that the 'expired' permit had become a deemed water right.

Seizing on the lacuna between the expiry of the permit in February and the coming into force of the new law in April, Williams conceived the legal basis for a private prosecution of the council. In short, if the

After completing his LLB Honours degree, and being admitted to the bar in 1972, **Andrew Brown KC** joined Russell McVeagh as a junior lawyer. In 1973 he was a Rhodes Scholar at Oxford, completing a Bachelor of Civil Law degree. Brown subsequently returned to Russell McVeagh and became a litigation partner in 1976.

Brown left the firm in 1988 to become a barrister and was made a Queen's Counsel in 2002. Brown went on to establish a highly successful legal career and he is currently recognised as the leading intellectual property barrister in New Zealand.

In 1970, David Williams employed Brown as his first summer law clerk at Russell McVeagh. This was when Williams was contemplating establishing EDS, so Brown became involved in the initiative. Brown was an inaugural EDS student director, and he remained on the board until 1973, when he left New Zealand to study at Oxford. As well as being involved in litigation on behalf of EDS and writing submissions for the society, Brown helped put together EDS's first promotional brochure and edited the early editions of *EDS News*.

RAEWYN PEART

permit had expired before the new law came into force, it was defunct and could not be deemed a water right under the new provisions. The council was therefore illegally discharging sewage into the Waikato River.

Williams filed proceedings in the Magistrates Court (now called the District Court) in May 1972. The proceedings were lodged in Williams's own name, as he did not think EDS would be given standing to appear in court. This was a risky move, as he could be held personally liable for court costs if the case failed, so Williams took care to explain to the judge that it was a public interest case which he was taking on behalf of EDS to draw attention to pollution of the Waikato River.

The case was novel. As reported in *The Dominion* in July 1972, 'The action is believed to be the first of its kind relating to the environment brought by a private group.'

Emeritus Professor Dick Bellamy, who would later join the EDS board and go on to be the society's longest-serving director, remembers reading about the case in the newspaper and thinking 'wow, finally this is doing something as opposed to just talking about it'.

Williams was represented by his old law-school friend Graham Hubble. Hubble managed to persuade the court that the allegations against Huntly Borough Council had a strong legal foundation. The judge found that the council had breached the law in two out of the three alleged instances. He dismissed the third allegation and discharged the council without conviction on the other two.

Not satisfied with the outcome, Williams and EDS appealed the finding on the third allegation to the Supreme Court (now called the High Court). In a decision issued by Justice Mahon in April 1973, the council was found to have breached the law in the first instance, and the matter was remitted back to the Magistrates Court for sentence. As Williams recalls, 'one of my memories of the case is a TV interview with the Mayor of Huntly Borough who, when questioned about the raw sewage being discharged into the Waikato River, made the wonderfully ambiguous statement – "don't worry it is 100 per cent"'.

The council then appealed to the Court of Appeal, at which point Paul Temm QC joined Williams's legal team. The court confirmed that the council had infringed the law. But, in the decision released on 1 March 1974, the judges discharged the council without conviction as it had since obtained the necessary permit to discharge raw sewage. Williams was awarded $250 in costs.

Although escaping punishment, the case motivated the Huntly Borough Council to take well-overdue steps to install a sewage treatment plant. The legal ruling also made it clear that others with temporary discharge permits, thought to number up to 300 at the time, would now need to apply for new water rights. It sent a strong message that blatant pollution of the Waikato River would no longer be tolerated.

ANOTHER OF EDS'S early cases focused on the control (or lack thereof) of the use of the pesticide 2,4,5-T. The weedkiller had first become available in New Zealand during the 1940s, and by the 1970s was being liberally used throughout the country to control gorse and other weeds as well as clear land for forestry. Aerial spraying was a common method of application in rural areas, where many families collected drinking water from their roofs. 2,4,5-T was also being marketed for use in home gardens and orchardists were spraying it on apricots as a growth regulator to increase the size of the fruit. It was one of the two ingredients that made up Agent Orange, the chemical used by US troops during the Vietnam War to destroy the enemy's forest cover and crops.

Questions around the safety of 2,4,5-T started to bubble up during the late 1960s, when overseas tests found that it caused birth defects in laboratory animals. Reports also emerged of increased human birth defects where Agent Orange had been used in Vietnam and amongst the families of returned servicemen.

The human toxin was not 2,4,5-T itself, but the small amount of dioxin (TCDD) that the weedkiller contained as a result of the manufacturing process. At that time, dioxin was one of the most powerful poisons known to science. By 1970, such evidence prompted the US federal government to ban the aerial spraying of 2,4,5-T in Vietnam and also to prohibit its use around homes and on most crops in the United States itself.

The largely uncontrolled and widespread use of 2,4,5-T continued unabated in New Zealand, however. In 1971 alone, over a million litres of 2,4,5-T concentrate was sold in the country. The protection of public health from the use of such a toxic chemical was ostensibly in the hands of the Agricultural Chemicals Board. The board was a government quango, established under the Agricultural Chemicals Act 1959, and was heavily stacked in favour of the industry. In particular, the Agricultural Chemical Manufacturers' Federation had a seat on the board, as did Federated Farmers of New Zealand (now Federated Farmers).

University of Auckland biochemist Dr Bob Mann had opposed the use of Agent Orange by the United States in Vietnam while a student at the University of California's Berkeley campus.

The Waikato River flowing through Huntly in 1961. This was the location of EDS's first court case, which sought to stop the discharge of raw sewage from Huntly township into the Waikato River.
WHITES AVIATION, ALEXANDER TURNBULL LIBRARY

His biochemical training meant that he was well aware of the risks to human health posed by dioxin. When Mann moved back to New Zealand to take up an academic position at the University of Auckland, he became concerned about the ongoing liberal use of 2,4,5-T in his home country.

In January 1972, two doctors from Te Awamutu published a letter in the *New Zealand Medical Journal* stating that drinking water contaminated by 2,4,5-T might have been a factor in the birth of two deformed babies. The mothers lived in houses where drinking water was collected from the roof and had been pregnant when aerial spraying was undertaken. The matter soon reached the press and public notice.

Seizing the opportunity that this increased profile of the issue created, EDS lodged a lengthy petition with the Agricultural Chemicals Board urging it take immediate action to ban the use of 2,4,5-T or, at the very least, impose greater restrictions on its use. The petition, worked up jointly by David Williams and Bob Mann, noted that 'one can at present buy from the food department of Farmers Trading Company in Auckland a bottle of 2,4,5-T which is not even labelled "poison" and bears no warning of possible danger to health'.

At the same time, Mann published an article in the *Soil and Health Journal* titled '2,4,5-T: An unregulated danger' in which he outlined the risks posed by the chemical's use.

Ivon Watkins-Dow Limited Paritutu Plant in New Plymouth during the 1960s, where 2,4,5 T was manufactured. EDS sought a ban or stronger controls on the use of the toxic chemical which was aerially sprayed over rural properties for weed control and was potentially associated with deformed babies. PUKE ARIKI, PHO2014-0125

> The World Health Organisation considers that to calculate the safe dose for humans, one should take the maximum safe dose in animals and divide by two thousand. Applying this safety factor, which is reasonable in view of the fact that we care about even one unnecessarily-deformed baby, the most dioxin a pregnant woman should take is around 0.004 millionths of a gram.
>
> How serious is the risk from 2,4,5-T? Well, even at the lowest dose quoted by the Department of Agriculture for killing gorse, about 4 lb/acre, accidental drift onto a farmer's roof would deposit about 90 millionths of a gram, using the manufacturer's own figures for dioxin content. Many farm families *drink* the water off their roofs. Even if the 90 millionths of a gram of dioxin were diluted in a thousand gallons of tank water, the farmer's wife would still take in enough dioxin in a couple of days' drinking to expect a malformed baby.

In response to such concerns, the Agricultural Chemicals Board appointed a sub-committee to consider whether further restrictions should be placed on the use of 2,4,5-T. The sub-committee did not call for any public input, but it did consult with Ivon Watkins-Dow, which had been manufacturing 2,4,5-T from a plant in New Plymouth since 1962.

The sub-committee found the science linking 2,4,5-T to birth defects inconclusive. Even so, it recommended a range of additional controls including a ten-fold reduction in the level of dioxin contained in the product, withdrawal of packaging aimed at the home market and placement of conspicuous warnings on labels. It also recommended that the aerial application of 2,4,5-T be 'discouraged' and not be undertaken within one mile of a homestead or urban area, which was the distance that spray was known to drift.

When considering the sub-committee's recommendations, the Agricultural Chemicals Board was far more reluctant to make changes that would impact agricultural users. Although adopting many of the recommendations, it declined to ban aerial spraying in the vicinity of people's houses on the basis that this would 'deny the benefits of applying 2,4,5-T from the air to the great majority of farmers'. Even then, the board failed to enforce its new rules, with household packs of 2,4,5-T still freely available in many gardening shops and lacking any warning labels.

Frustrated by the lack of action, EDS decided to launch legal proceedings. These took the form of a 'writ of mandamus', an ancient proceeding available under common law to enforce the undertaking of a public duty. Russell McVeagh lodged the writ, on behalf of EDS, in the Wellington Supreme Court (now the High Court) on 15 September 1972. David Williams represented EDS at the subsequent hearing, in front of Justice Haslam, supported by Andrew Brown who was now working at Russell McVeagh as a junior lawyer.

Williams can still remember the hearing clearly.

> At one point in the proceedings Justice Haslam said to me 'Mr Williams there is something distinctive about you — you are the only person in Court who is not a Rhodes scholar.' The judge, Don Mathieson, who was acting for the Crown, and my junior Andrew Brown were indeed all Rhodes scholars. This was not calculated to improve one's sense of confidence. But I quickly replied, 'at least my middle name is Rhodes'.

The Court ultimately ruled that EDS did not have standing to bring the suit, which was disappointing for the young EDS lawyers. But a message was sent to the government. Decisions affecting public environmental health would come under scrutiny.

It took another 14 years for the use of 2,4,5-T to be banned in New Zealand. By that time, there was concern that residents close to the New Plymouth manufacturing plant had been exposed to high levels of dioxin. Later studies showed this to be the case, with those living in the area during the 25 years of manufacture having significantly elevated levels of dioxin in their bodies. This was thought to have resulted from residents breathing contaminated air as well as eating homegrown leafy vegetables and soft-skinned fruit on which dioxin had settled.

Reflecting on the case some 50 years later, Andrew Brown KC observed,

> it was very exciting. No-one in New Zealand had done anything like that before. It took brave steps by David to conceive of the legal action and be prepared to take it on. At that stage, Russell McVeagh had become quite an activist partnership compared to the rest of the legal profession, and was prepared to go along with it. So that was exciting as well.
>
> We were hopeful that the case would succeed, but the judge was very conservative and administrative law was in its infancy. We got ruled out on the grounds of not having standing to sue which was very frustrating. But subsequent history with Ivon Watkins-Dow in New Plymouth showed how right the case was. The contamination of the site, and the people nearby who were affected by deformity in their children, was a national disgrace. Bob Mann was absolutely right!

—000—

Having cut its teeth on the prosecution of Huntly Borough Council and the 2,4,5-T legal challenge, EDS was ready for bigger things. The need for EDS's expertise soon arose; again in the context of the Waikato River.

First announced by government in July 1972, the Huntly Power Station was to be the largest thermal power scheme ever built in New Zealand. It was part of the government's push to significantly increase power generation in the North island, a broader controversy that EDS became closely involved in (as described in the next chapter).

The small town of Huntly became the preferred location for the mammoth plant. It had significant coal deposits which could fuel the power station, was located on the banks of the Waikato River where cooling water could be sourced, and was relatively close to Auckland and other population centres where the power would be used. It was also thought that the residents of the small and declining coal-mining town might be more receptive to the enormous development than those living in other prospective locations.

The proposed site for the station was just to the north of the small Huntly township, on the west bank of the river. Immediately to the south was the Waahi Marae, home to the Ngāti Mahuta people, including the Māori Queen Dame Te Atairangikaahu.

Despite the power station structure being over 400 metres long and having boiler houses 71 metres high, plus two enormous chimneys reaching 150 metres into the sky, the government did not need planning consent to build it. This was because, at that time, the Crown was exempt from any controls under the Town and Country Planning Act 1953.

But the government did require a permit to take and discharge river water for cooling purposes due to new procedures recently introduced into the Water and Soil Conservation Act 1967. These required an application to be made to the National Water and Soil Conservation Authority, with those directly affected by the proposal able to appeal the authority's decision to the Town and Country Planning Appeal Board. Such environmental scrutiny was novel for government, with the Huntly Power Station being one of the first government projects to go through the new process.

The New Zealand Electricity Department applied for the water right in early 1973. This sought permission to extract up to 34,200 tons of river water per hour (27 per cent of the river's flow at Huntly), and then to return it to the river. The inherent inefficiency of generating electricity through burning coal meant that for every 1000 megawatts of energy produced by the power station, an equivalent of 1275 megawatts would be produced as waste heat and this needed to be discharged by some means. The river water, once it had been through the power station, would be heated by up to 8 degrees centigrade.

In order to minimise costs, the department had opted to avoid the extra expense of installing cooling towers or ponds, which could remove excess heat from the water prior to discharge. Instead, it was hoping that the heated water would readily mix with, and dissipate within, the Waikato River's natural flow. This was somewhat optimistic, given that it proposed to extract over a quarter of the river's flow and return it at a significantly elevated temperature.

On lodging the water permit application, the government's engineers claimed that the

Huntly Power Station under construction in 1979. Depsite the structure being over 400 metres long, having boiler houses 71 metres high and two enormous chimneys, the government did not need planning consent to build it.
WHITES AVIATION, ALEXANDER TURNBULL LIBRARY

water take and discharge would have only minor environmental effects on the river. This was a bold assertion, as they had not carried out any research into possible impacts on aquatic plants, fish or other river life. But despite this considerable gap in information, consent to the proposal was granted by the National Water and Soil Conservation Authority in June 1973, an unsurprising decision given that the government was promoting the scheme and the authority was chaired by the Minister of Works Percy Allen.

EDS lodged one of only two appeals to the Town and Country Planning Appeal Board contesting the decision. In its appeal, the society asserted that the water rights should not have been granted at all, as the potential effects of discharged heated water on the river's ecology were unknown and it was concerned that such a large volume of heated water would adversely impact aquatic life. In addition, due to the river's already polluted state, an increase in temperature could result in algal blooms. EDS did not oppose the building of the Huntly Power Station itself but sought much tighter conditions on the grant of water rights.

Planner and EDS director since 1999 Graeme Lawrence recalls,

> I was working in a voluntary capacity with Robert Mahuta at the Waikato University Māori Research Centre. We were trying to bring Māori cultural values and aspirations, and their relationship

The Waikato River, Huntly township and Huntly Power Station. The water permit that government sought for the power station proposed to extract over a quarter of the river's flow for cooling water and to discharge it at a significantly elevated temperature. EDS sought much stricter controls over the permit in order to protect river life. ROB SUISTED/NATURESPIC

with taonga, into the arena of the Town and Country Planning Appeal Board. Then this warrior 'knight in shining armour' appeared in the form of David Williams. EDS was framing issues around the Huntly Power Station much more narrowly than we were. The focus on ecology of the river was amazingly close in Pākehā scientific terms to one part of what the Māori community was trying to articulate in terms of mātauranga Māori — a wider more inclusive understanding of the values of the Waikato River. They ended up as partners in the preliminary hearings.

Demonstrating either supreme confidence in achieving a positive outcome or simply contempt for the legal process, the government commenced construction of the power station in 1973 prior to the two appeals being heard.

Securing technical evidence to support EDS's case was left in the hands of University of Auckland microbiologist Dick Bellamy. Bellamy had spent some years in the United States undertaking post-doctoral research and had taken the opportunity to travel widely around the continent. He had visited national parks as well as the industrial heartland. He saw 'big wetlands used as rubbish dumps, huge mountains of smouldering rubbish, and oil refinery after oil refinery pouring out all sorts of pollution'. He also found local beaches fenced off to everyone but local residents. 'It got me wound up,' he recalls.

Bellamy settled back in Auckland in 1968, working as a virologist at the University of Auckland. He was fired-up about environmental despoliation but there was little in the way of activism on the issue in New Zealand. Around a year or so after EDS was established, Bellamy saw a notice in the Department of Biological Sciences library inviting people to become involved in the society. He liked the idea of using the law to achieve environmental outcomes and started attending EDS meetings.

As Bellamy explained,

> we had no money. We had a team of lawyers but few scientists. We had a meeting every month or so in David's office. But these came to an abrupt end when we started making submissions on things. The Harbour Board, which was Russell McVeagh's biggest client, discovered that the EDS post-office box was the same as Russell McVeagh's. David's partners thought we should move on, so we went elsewhere.

EDS director Graeme Lawrence on a society field trip to Ngunguru Sandspit. Prior to becoming a planning consultant and joining the EDS board in 1999, Lawrence was chief planner and then manager of environmental planning at the Thames-Coromandel District Council. RAEWYN PEART

Bellamy got drawn into the Huntly Power Station case.

> David Williams organised a meeting with the Electricity Department honchos in his office at Russell McVeagh. They were used to dealing with 'greenies' who met in some backyard sleepout. But this time they had to come into Queen Street and go up to the 10th floor of a smart office block where there was a view out over the harbour.
>
> They explained to us what a marvellous idea the power station was and how the nation needed it. Towards the end of the meeting they asked, 'well are you happy now, will you withdraw your objection?'. David replied diplomatically 'very good talking to you and exchanging useful information. We will now have to consult and think about it.' We waited a week and then advised them 'now we have considered the matter, we are proceeding'.
>
> I had the job of organising the scientific side of the case. I didn't know anything about the topic. I was totally out of my depth. At that time, I was going to a major scientific conference in the US once a year. I discovered there was a guy in Washington DC who had just run a big case on the effects of a thermal power station on a river and I went to see him. I learnt that it was not just the immediate effect of the hot water that was an issue, there were

> other things to consider. For example, once you put hot water into a river, the fish congregate around the outlet. Every so often the engineers clean out the fouling in the cooling pipes by adding hypochlorite and the fish in the thermal plume in the river die. Then there is the question of how the fish get past the thermal barrier when migrating up the river.

To assist with the case, Bellamy sought out his old primary school friend and University of Auckland colleague Dr Mike O'Sullivan. O'Sullivan's speciality was engineering science and he was an expert computer modeller. He developed a computer simulation to show how the heated water would likely behave once it was discharged into the river's flow. This demonstrated that the assertions of the government engineers were way off the mark. Rather than the heated water becoming well-mixed within 800 metres of the outfall, as claimed by them, Mike's modelling showed that the affected area would extend up to 3 kilometres downriver.

But for EDS to establish its case for tighter conditions on the water permit, it needed to demonstrate that this extended mixing zone would have negative impacts on river life. For this, Bellamy approached Dr John Leader, an English-trained freshwater biologist who at that time was working at the University of Auckland. Leader was able to describe how the ecology of the river would likely be impacted by the dramatic changes in temperature shown in O'Sullivan's modelling.

In September 1973, Russell McVeagh lawyers David Williams and Andrew Brown presented the EDS case to the Town and Country Planning Appeal Board, chaired by Arnold Turner. Despite the legal cogency and scientific strength of EDS's case, the Appeal Board found that the society did not have standing to appeal, as it was not 'a person which or who claims to be detrimentally affected by the decision of the authority'. EDS's appeal was dismissed on the basis that the society had no status to bring it.

Huntly Power Station in 2023. The power station was finally commissioned in 1983 with tight conditions on its water rights due to submissions by EDS and Robert Mahuta. This restricted its operation on hot summer days in order to minimise impacts on river life. A cooling tower was eventually built in 2004. RAEWYN PEART

But all was not lost. There was the second appeal by Robert Mahuta. To support the appeal, Mahuta gave evidence on Māori concerns and values and his iwi's cultural connection with the river. The Appeal Board did not find such cultural matters relevant under the Water and Soil Conservation Act and much

Professor Richard (Dick) Bellamy CNZM completed degrees in botany at the University of Auckland before undertaking a PhD in microbiology based at the DSIR research station in Mount Albert. It was there that he first became interested in viruses. Bellamy then undertook post-doctoral studies at the Albert Einstein College of Medicine in New York.

When Bellamy came back to Auckland in 1968, he established a genetics group at the University of Auckland. He first became actively involved in EDS through the Huntly Power Station case. As he recalls, 'at that stage people thought I was doing no work for the university and was spending all my time on environmental issues. But you could get away with doing public interest work then much more than now, so long as your publication record didn't drop off and you did your teaching. These days it would be impossible to be the coordinator of science for a big case like that.'

Bellamy brought a very useful set of complementary scientific skills to the small group of EDS lawyers. Andrew Brown remembers him as being 'quite a towering figure. He was very highly regarded and he always spoke authoritatively and sensibly.'

Retired Court of Appeal judge, Tony Randerson KC, recalls 'Dick Bellamy used to ring me up at 8am to have long conversations. He'd ask, "I've just had an idea, do you think it will run?" He had such a creative mind and was an incredible lobbyist. He used to write to a Minister of the Crown and copy it to all the other ministers and the local council, etc. Then each would reply and copy in all the others. Before long, in various bureaucracies, there was a pile of paper building up. Then they would think "the file is so big it must be an important issue". It was a very clever technique with minimal effort to achieve large files in different people's offices. He was a very smart guy.'

Bellamy became the Founding Director of the School of Biological Sciences in 2000 and subsequently became Dean of Science. In 2005 he was awarded a CNZM for services to science and education. On retirement in 2008 he became an Emeritus Professor. Bellamy joined the EDS Board in 1973, and was still on the society's board in 2023, making him the longest-serving board member.

DAVID HARTMAN

of his evidence was dismissed as irrelevant due to a lack of specificity on the water quality effects of the power station. However, it was clear that the Waahi Marae, being next door to the power station site and a long-standing user of the river, was affected by the development and that Mahuta and the people he represented had standing to appeal.

The Appeal Board was sympathetic to the issues EDS had raised and it held that, provided one appellant could establish status to appeal, 'all questions raised by the application are open to review'. Through this legal nicety, the Appeal Board was able to base its decision to tighten up consent conditions on the society's evidence, despite dismissing EDS's appeal. And although much of the evidence given to support Robert Mahuta's appeal was dismissed, some of the river health safeguards he was seeking were achieved through the EDS case. In this way, the two appeals and the supportive relationship between EDS and Ngāti Mahuta enabled both parties to jointly achieve a positive outcome. Taking Charles Wurster's advice to sue the government even if EDS had no standing was starting to pay dividends.

Significantly, the New Zealand Electricity Department agreed to carry out a full biological survey of the river prior to commissioning the power station, to monitor the effects of cooling water against this baseline, and to make the results public. This helped establish a precedent that environmental studies should be carried out when applying for water rights.

The decision was also important for another reason. It was the first time a court had treated the Crown in the same way as private citizens under the Water and Soil Conservation Act. Previously, the Crown had asserted that lower environmental standards should apply to its applications for water rights.

The lack of acknowledgement of Māori values in the case also helped highlight an issue which was addressed when a reformed Town and Country Planning Act was passed four years later, in 1977. 'The relationship of the Māori people and their culture and traditions with their ancestral land' was included as one of the matters of national importance to be 'in particular recognised and provided for'. As Graeme Lawrence observed,

> at that time, 'Save Manapouri' was the environmental thing. But in terms of lasting influence on our environmental legislation, the Huntly Power Station loomed large. The Huntly Power Station case was a trigger point for a lot of things, including contesting science through bringing scientific evidence. And where the legislation was found inadequate, it was later amended.

When comparing the controversy over the establishment of the Huntly Power Station with the Manapōuri campaign, historian Jo Whittle concluded that 'events at Huntly had a considerable influence on the development of environmental policy and therefore merit inclusion in the history of the emergence of environmentalism in New Zealand'.

The Huntly Power Station was finally commissioned in 1983. The tight conditions on its water rights, due to the EDS and Mahuta submissions, meant that it could not run at full capacity on hot summer days as the maximum allowable temperature of the discharge water would be breached. When the power station was subsequently expanded in 2004, a cooling tower was built. Today, in a carbon-constrained world, the power station is facing decommissioning in favour of renewable energy.

2. Power play

A MAJOR CONTROVERSY during the 1970s, and one that the EDS became deeply embroiled in, was how and to what extent New Zealand should generate electricity to service a growing population and economy. The issue was of considerable significance to the natural environment because of the risks and impacts created by the construction and operation of large power-generation plants.

IN THE EARLY 1970s, even nuclear power was on the table. In 1972, Electricity Department General Manager E. B. Mackenzie announced that he expected his department to be working on plans for nuclear stations within 10 years. Costed plans had already been produced for a heavy water nuclear plant on the banks of the Waikato River. As early as 1965, Rodney Member of Parliament William John Scott had referred to a proposed site for a nuclear power station in the Kaipara area. There were also plans to build a series of thermal power stations in Tāmaki Makaurau Auckland in order to burn the copious quan-tities of natural gas recently discovered off the Taranaki coast.

Much of this activity was fuelled by the fore-casts of overzealous government energy planners. They predicted that the rapid growth in power usage experienced immediately after the Second World War, when much of the country was being electrified, would carry on into the future unabated. Also the 1973 oil crisis (when OPEC, the Organization of Petroleum Exporting Countries, reduced production and put an oil embargo on countries that supported Israel in the 1973 Arab-Israeli War) saw oil prices soar and prompted the government to focus on domestic sources of energy to help insulate the country from future fluctuations in overseas energy supplies.

Despite the significance of electricity generation decisions for the future of the country, there had been no public consultation or debate about the options, and the basis for the optimistic power predictions remained buried behind the impenetrable wall of the Official Secrets Act 1951.

Matters came to a head when the government announced its intention to construct a thermal power station (called 'Auckland Thermal No. 1' — as a 'No. 2' was also being planned) at Te Atatū on the shores of the Waitematā Harbour. Local residents were outraged. EDS director Dick Bellamy recalls attending a large public meeting on the proposal in October 1973. 'There were so many folk in the Te Atatu school hall that people were sticking their heads through the windows to hear what was going on inside.' The local opposition was so intense that Norman Kirk's Labour Government was persuaded to abandon the site (although not the project) in late 1973.

It was in this context that EDS decided to investigate future energy provision in New Zealand. The initiative was led by a group of University of Auckland scientists, including Bob Mann, Dick Bellamy and Mike O'Sullivan. They wanted to ensure that government energy decisions, especially any decision to go nuclear, would only be made after full public consideration of the options. As *From Manapouri to Aramoana* book author Roger Wilson noted, 'Mann quickly established himself as a forceful and well-researched opponent of nuclear power, forcing the NZED [Electricity Department] to take steps to ensure that their staff kept very tight security on the moves they were taking to investigate the subject.'

One of the early initiatives of the EDS group of scientists was to convene the New Zealand Energy Conference, which took place in Auckland in late May 1974. It was designed to open up energy matters to wider debate. EDS invited a series of overseas speakers, including Dr Ernst R. Habicht Jr, who had been working on energy issues for the US-based Environmental Defense Fund (EDF). The conference was opened by Prime Minister Norman Kirk and closed by Minister of Energy Resources Warren Freer. At the end of the

---

Previous pages: Lake Benmore hydro power works.
RAEWYN PEART

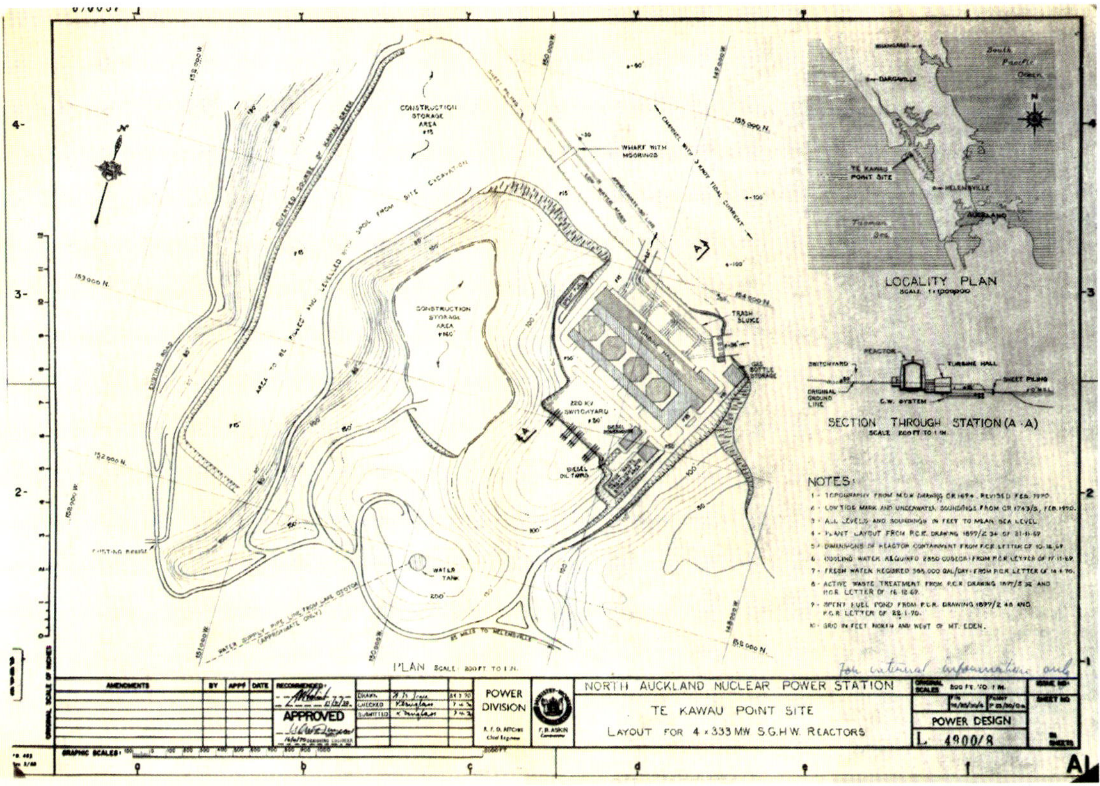

Plan for a nuclear power station at Te Kawau Point, Kaipara Harbour, 1970. The government had been progressing plans to develop nuclear power stations in New Zealand from the mid-1960s without any public debate as to whether nuclear power was in the best interests of the country. ARCHIVES NEW ZEALAND

conference, EDS called for a full public inquiry before any decision was made by government to establish nuclear power plants in New Zealand.

The event was so successful that three more energy conferences were held in subsequent years. As Mike O'Sullivan recalls,

> they were the first conferences like that in New Zealand. We got lots of government folk to come along. We also secured sponsorship to invite overseas speakers. We gained exposure for the whole energy planning issue, which had a big influence on politicians. They were well-attended and good conferences.
>
> At a meeting in Pukekohe on the Auckland Thermal No. 1 proposal, I remember local community leader Noelene McLarin waving the proceedings from the first New Zealand Energy Conference at the Electricity Department person and saying 'have you read this book?' So it was quite effective.

ooo

DURING THE 1970s, a government Committee to Review Power Requirements produced an annual annexure to the Parliamentary Debates that projected the country's power requirements. Its 1974 report cautioned that New Zealanders would probably need to accept nuclear power by 1990, with a decision required in 1975 or soon thereafter. The following year, the committee proposed that the starting date for nuclear power would need to be brought forward to 1979–80, with the building of twin 600-megawatt reactors at a cost of some $463 million.

But the committee had lost a considerable amount of credibility by this stage. It had forecast a 9.6 per cent increase in peak electricity demand for the 1973–74 year, but actual growth had been just 2.3 per cent. For the 1974–75 year it forecast even higher growth of 12.9 per cent, but demand had actually declined that year. Seemingly undeterred by these discrepancies, the committee forecast still higher growth for the 1975–76 year of 17.7 per cent.

In the longer term, the committee was projecting an annual increase in demand of 6.4 per cent. Building power stations to meet this demand would have required the government to commit a third of its total capital expenditure each year. Criticisms of the projections were dismissed by the Minister of Electricity Ron Bailey. 'It would be a good idea for uninformed people to leave the power production projections of New Zealand to experts,' he suggested.

When EDS director Mike O'Sullivan examined the government's power projection figures, he found them to be patently ridiculous. He estimated that it would take at least 21 years for peak demand to double, as opposed to the power planners' estimates of only a decade. To meet the government's projections, O'Sullivan calculated that the country would need to build a new 1000-megawatt power station every year by 1995, increasing steadily to two such stations each year by 2005. As set out in EDS's submission to the government:

> Even allowing for the introduction of nuclear power into NZ it is inconceivable that 20 large power stations can be built in NZ over the 15-year period 1990 to 2005. There is no prospect of finding suitable sites, finance to build them or fuel to run them. If viewed over a further 15 years from 2005 to 2020 the 6.7% growth rate becomes absurd requiring the building of 52 1000 MW power stations.

However, such anomalies in power projections did not dissuade the government from forging ahead with its power-station building programme, with both nuclear and thermal power stations actively on the agenda.

In 1975, Minister Bailey announced that six sites for the location of nuclear power stations were under consideration: two on the Firth of Thames and the others at Oyster Point on the shores of the Kaipara Harbour, Makara and Baring Head in Te Whanganui-a-Tara Wellington, and on Te Pātaka-o-Rākaihautū Banks Peninsula. EDS obtained a document from Electricity Department sources calling for the government to make an 'in principle' decision on nuclear power the following year. Environmental impact reports on the various sites would then be published and construction of the first reactor would commence in 1979. But the public had yet to be engaged in any debate about whether nuclear power was a good idea for New Zealand in the first place. It looked like the Electricity Department was planning to present a *fait accompli*.

EDS continued pushing for a full-scale public inquiry into the matter. Its efforts helped flame growing public disquiet about nuclear power. This prompted the Labour government to establish a small 'Fact Finding Group on Nuclear Power' to examine its possible

Professor Michael (Mike) O'Sullivan graduated from the University of Auckland with a Bachelor of Civil Engineering, a Bachelor of Science in mathematics and a Masters of Engineering in applied mechanics. He then completed a PhD at the California Institute of Technology. After a brief period at New York University, O'Sullivan returned to the University of Auckland to take up a position in the Department of Engineering Science. His research has focused on computer modelling of geothermal fields and other environmental and computational fluid dynamics problems. One of his projects focused on Mururoa Atoll. His modelling of temperature data from French sources indicated considerable leakage of radionuclides through the atoll's rock walls.

RAEWYN PEART

O'Sullivan became involved in EDS through his friendship with University of Auckland colleague Dick Bellamy. He was on the EDS Board from 1974 to 1987 and he contributed to many of the society's cases during that period including the Huntly Power Station water rights and the Waiau Pā thermal power station. He was particularly keen to see the government acquire more coastal land.

environmental consequences when compared with alternatives. However, the largely technical group deliberated behind closed doors, and the public was still excluded from the debate.

Sensing the public mood, in the campaign leading up to the 1975 General Election, the National Party promised not 'to introduce nuclear power generation in New Zealand until a public inquiry into all aspects of this source of energy has taken place and until it is convinced that technological aspects have been satisfactorily resolved'. National ultimately won the election and formed the new government under the leadership of Prime Minister Robert Muldoon.

To further raise the profile of the nuclear energy debate, EDS joined with other environmental groups to form the Campaign for Non-Nuclear Futures. The coalition decided to launch a public petition opposing the use of nuclear power reactors in New Zealand. It was highly successful, with 333,087 signatures presented to Parliament in November 1976. This made it, at the time, the largest petition ever circulated in New Zealand, substantially eclipsing the 1970 Save Manapōuri petition.*

By September 1976, the government had announced the establishment of the promised Royal Commission of Inquiry into Nuclear Power. It was chaired by Sir Thaddeus McCarthy, a former President of the Court of Appeal. Rather than primarily being a scientific review, the inquiry was focused on enabling the public to be heard on the issue. University of Auckland economist and then EDS director Professor Robin Court gave evidence to the inquiry on behalf of EDS and others. He questioned the Electricity Department forecasts, suggesting

* Up to that time, it was the biggest petition New Zealand had ever seen, with 264,907 signatures opposing raising Lake Manapōuri.

that they could not be considered forecasting at all but merely 'self-fulfilling plans'. He also challenged the government's costings associated with nuclear power plants. In his view, establishing nuclear power in New Zealand would be hugely expensive.

In the Royal Commission's final report, Professor Court's evidence on likely nuclear power costs was noted as most helpful on that point. The report stated that:

> His evidence received wide publicity, and was directly or indirectly used as a reference source in many other submissions. His professional status had a bearing on this, as had his basic contention that the capital cost of the NZED [Electricity Department] project, if started now and completed around 1984 would amount to a figure in excess of $2 billion ... While some of Professor Court's conclusions were challenged, particularly in cross-examination, we acknowledge that he caused the NZED to undertake a major clarification of the costs submitted in its background paper.

During the hearing, the Electricity Department had been forced to admit that the costs of building a nuclear power plant had risen considerably over the previous two years, doubling from its initial $463 million estimate to some $924 million. The revised figures indicated that the cost of nuclear power would be around three cents per unit, considerably more costly than coal-generated energy, and almost as expensive as electricity generated from oil.

When it came to the risks of nuclear power, a representative of the Department of Scientific and Industrial Research (DSIR) was reported as asserting under cross-examination that it was better for the public to be kept ignorant of some nuclear problems to avoid causing alarm. 'Some things are best made known in retrospect after they have been put right,' he contended. 'We are building a technological society in which there will inevitably be more restrictions on civil liberties.'

The Royal Commission's deliberations carried on throughout 1977. But by this time, the impetus was running out of the Electricity Department's push for nuclear power. The energy forecasts had been 'revised', and they indicated that nuclear power would not be needed nearly as early as previously thought. In addition, the department gradually accepted that if there was to be a nuclear power programme in New Zealand there needed to be a large measure of public acceptance that it was in the national interest.

When the Royal Commission finally tabled its report in Parliament, in May 1978, it confirmed the lack of an urgent need to make a decision on nuclear power. The decision could be left until the 1992–96 period. In the meantime, the commission somewhat diplomatically advised that it would be prudent to plan development in such a way that future energy demands could be met from renewable rather than finite sources. Although the Royal Commission did not fully exclude the prospect of nuclear power in the future, for the moment at least, the issue was buried. It never reappeared.

As Roger Wilson observed,

> Government had underestimated the thoroughness of research and the professionalism displayed by the environmental movement, and had also underestimated the depth of feeling amongst the public over the question of nuclear power.

Dr Robert (Bob) Mann completed a Bachelor of Science in chemistry from the University of New Zealand and then a Masters in Science (Hons) from Victoria University of Wellington. He then headed overseas to complete a doctorate at the University of California Berkeley. Mann returned to New Zealand to take up an academic position at the University of Auckland, becoming a senior lecturer in biochemistry and then a senior lecturer in environmental studies in the Department of Town Planning. He served for eight years on the Environmental Council Working Party on Energy and 11 years on the Toxic Substances Board.

Mann became a board member of EDS in 1972 and served until 1983. He was a major driver behind EDS's 2,4,5-T and nuclear power campaigns as well as its later work on LPG installations. He also undertook valuable work to promote more sustainable onsite wastewater systems. Andrew Brown remembers Bob Mann from the early years of EDS as 'a cool guy, really earnest and convincing and absolutely passionate'.

Graeme Lawrence as a practising planner in the Thames Valley United Council area, prior to the 1987 local government reforms, remembers Bob Mann's efforts to demonstrate there was a better way to provide onsite wastewater systems. Mann advocated designing systems that were appropriate to site conditions. His methodology became accepted as a means of replacing the outdated and environmentally unsound traditional septic tank systems in some bach settlements without having to resort to full reticulation.

Barrister Richard Brabant, a former EDS director who worked closely with Mann on the LPG cases, recalls him as being very bright but unusual. 'He was his own man in a very definite way. He didn't fear anyone and he didn't have any respect in the traditional sense for authority.' When Brabant was calling Mann as a witness in front of the Planning Tribunal he faced the challenge of what Mann would wear. Mann had been known to turn up to court in sandals, something which the presiding judge (Judge Sheppard) did not view favourably. As Brabant recounts, 'I worked on Bob for ages as best I could. I was pretty successful. He came along in slacks and a jacket. He wouldn't do a tie but I got shoes on his feet.'

For environmentalists, the debate over nuclear power was a landmark. Not only had they organised nationwide on a single, bitterly fought issue, but they had been able to maintain the presence of several of their 'experts' at the hearing of the Royal Commission for several months. The campaign also represented one of the first instances where the environmental movement was forced into a highly technical debate. It was a trend that continued; in the later 1970s and early 1980s environmental arguments were to become increasingly complex and technical.

AT THE SAME TIME as the nuclear power issue was playing out in New Zealand, the Electricity Department was also progressing plans to build a gas-fired, steam-powered electricity plant in Auckland (Auckland Thermal No. 1). In 1974, the department had released a preliminary environmental impact report for six potential sites: Oyster Point on the Kaipara Harbour, Muriwai Beach, Whatipū and Wattle Bay on the Manukau Heads, Waiau Pā on the southern shores of the Manukau Harbour and the Ōrere Point–Kaiaua coast on the shores of the Firth of Thames.

This aerial view of Clarks Beach, Manukau, Auckland, in 1986 shows the wide sweep of the bay near the top of the image, above the Clarks Beach settlement. This is where the government proposed to construct large cooling ponds for the Auckland Thermal No. 1 power station. WHITES AVIATION, ALEXANDER TURNBULL LIBRARY

EDS did not consider any of the proposed sites to be acceptable due to their environmental sensitivities, particularly if large cooling ponds were to be constructed over the intertidal area. Instead, the society suggested that the power station (if it was to be built at all) should be located in Mangere, next to the wastewater treatment plant. This could enable a 'win-win' arrangement, whereby the treatment plant's oxidation ponds would provide cooling water for the power station and the additional heat from the power plant would significantly increase the filtering efficiency of micro-organisms within the ponds. However, applying such an industrial ecology approach to waste streams was way ahead of its time.

The end-use of waste heat from the power plant was a significant issue, because it was estimated that around 2000 megawatts of heat would be produced for every 1400 megawatts of electricity that the power station would actually generate — highlighting the gross inefficiency of using gas for the proposed method of generating electricity.

The government ultimately opted for the Waiau Pā site on the south-western shores of the Manukau Harbour. The power station was to consist of four 350-megawatt units, which would be fuelled primarily by natural gas from the Maui Gas Field, off the coast of Taranaki, but with capacity to use oil as a standby fuel.

One of the most controversial elements of the proposal was the enormous cooling ponds which were to extend over 560 hectares of intertidal waters, an area the size of some 800 rugby fields. The use of ponds instead of cooling towers was due to their lower cost. There was a suggestion that the ponds could be used for rowing and yachting, with such recreational possibilities seen as a good 'selling point' to gain public support for the power plant.

The intertidal area concerned was no wasteland. A study of the ecological impact of the power station proposal highlighted that the area constituted 'rich sheltered harbour habitat ... providing a rich food source for large populations of fish and birds'. In particular, the seabed had exceptionally high densities of small Crustacea, which was an important food source for juvenile fish.

Residents from the primarily rural farming community quickly mobilised to oppose the proposal. This led to the establishment of the Manukau Harbour Protection Society in early 1975. Local Māori associated with the nearby Whatapaka Marae were also very concerned about the potential impact of the power station, particularly on their customary fishing areas. The Manukau Harbour had long been an important food source for Waikato Māori and there was concern that 'the site proposed for the power station has placed in jeopardy our sacred areas and traditional fishing rights'. All this opposition came as a surprise to the Electricity Department, which was unused to such intense political and environmental pressure.

Led by local resident Noeline McLarin, the Manukau Harbour Protection Society persuaded the Franklin County Council to hold informal hearings on the proposal. They were only to be 'informal' due to claims by the Electricity Department that, as the Crown was not bound by the Town and Country Planning Act 1953, it did not need to apply for a scheme change in order to rezone the Rural A site.

Along with many others, EDS presented oral and written submissions at the hearing, which was held in November 1975. The council ultimately decided that the power station should be located on open coastline rather than on the shores of the Manukau Harbour. But the decision had no legal weight and government continued pursuing the harbour option.

Just prior to the hearing, the Treaty of Waitangi Act 1975 had come into force. It established the Waitangi Tribunal and gave it jurisdiction to consider claims of Treaty breaches by government or any state-controlled body occurring after 1975. In February 1976, a deputation from the Waikato hapū met with Minister of Māori Affairs Duncan McIntyre at the Whātāpaka Marae. They asked him to constitute the Waitangi Tribunal to hear their concerns about Auckland Thermal No. 1. The minister agreed to do so, and two claims were eventually lodged with the tribunal. The claims alleged that the power station breached the Treaty guarantee that Waikato iwi had a right to continue to fish in the area.

The March 1976 edition of *EDS News* noted that 'The little known Treaty of Waitangi Act passed by Parliament suddenly assumes enormous significance in that it creates the machinery for protecting some of the traditional rights of the Maori people.' The article went on to highlight the potential benefits of Maori and environmentalists joining forces to oppose the project.

> The combining of forces between the local Maori people and environmental groups heralds a link that must continue to grow... Common ground between the two peoples of our nation and especially between the ecologically sound traditional Maori life style and the skill and expertise of environmental groups in dealing with complex legal, scientific and political matters, can contribute significantly to the conservation and wise use of our remaining natural resources.

Waikato hapū lodged a claim with the Waitangi Tribunal over the loss of their rich fishing grounds at Waiau Pā due to the proposed cooling ponds to be built as part of the Auckland Thermal No. 1 power station. EDS assisted with the case, which was the first environmental matter to be considered by the newly minted tribunal. RAEWYN PEART

EDS director Stephen Mills, who was at the time teaching at the University of Auckland Law School, turned his legal mind to the issue of whether a legal challenge could be mounted against the power station. On a close reading of the relevant statutes, he discovered a largely forgotten provision in the Clean Air Act 1972 that potentially required a government proposal requiring a permit under the Clean Air Act (which the power station did) to also comply with the Town and Country Planning Act. The provision had never come under scrutiny or been applied before. If Mills was right, the Waiau Pā proposal could be brought under the ambit of the Town and Country Appeals Board, and potentially higher courts on appeal.

When EDS first raised this contention, Prime Minister Bill Rowling responded that the Crown did not regard itself bound by the Town and Country Planning Act and would not voluntarily agree to observe it. So, in December 1975, EDS issued proceedings in the Wellington Supreme Court seeking a declaration that the Minister of Electricity needed to comply with the Act in building the Waiau Pā power station. Despite the Government's earlier protestations, when the matter was finally heard, the Crown conceded that the EDS legal argument was in fact correct and the court ruled accordingly.

In an article headlined 'Landmark Courtroom Triumph' in the August 1977 edition of *EDS News*, Mills observed, 'On a wider level the court action ought to be a lesson to government on the ability of the Society to use the law carefully to block projects which, on the basis of widespread public opinion and scientific and legal evidence, are ill-conceived and unacceptable.' The case sent a signal to government that, where the law permitted, EDS would hold it to account.

By this time, the Electricity Department had released a second environmental impact report focused on the Waiau Pa site. The minister had also served notice on the Franklin County Council that land would be requisitioned for the power station. At this point, the minister agreed to submit to normal town planning hearings on the proposal, although he still claimed that there was no legal obligation to abide by the outcome. It was the first time government had allowed a major public work to go through formal town planning procedures.

Franklin County Council convened a second hearing in early 1977 and heard 77 submissions. The Commission for the Environment and the Auckland Regional Authority broadly agreed with the concerns expressed by local Māori, EDS and the community. At the outset of the hearing, the Electricity Department asked the council for an adjournment, on the basis that it was now investigating a series of combined cycle plants* that might make the Waiau Pā

* A combined cycle plant means that the waste heat from the gas turbine is diverted to a steam turbine thereby providing up to 50 per cent more power from the same fuel than the traditional inefficient plant being proposed at Waiau Pā.

power station unnecessary. The council decided to press on regardless.

Water rights were also required for the power station, and the application for these was heard by the Auckland Regional Water Board at the same time as the second council hearing was progressing. EDS made submissions at this hearing as well. The Water Board issued its decision in April 1977. It found that the proposed cooling ponds would have unacceptable implications for the Manukau Harbour and adjacent land, and that without sufficient information on all relevant matters, the board could not recommend that a water right be granted.

Through EDS, lawyer Richard Brabant and freshwater ecologist Stella Penny assisted representatives of the Whātāpaka Marae to appear before the Waitangi Tribunal. One witness testified that the large cooling pond system 'would have a major adverse ecological effect upon the Manukau Harbour because it would require the destruction of a large area of inter-tidal habitat of very high ecological value'. It would also have meant the destruction of all edible fish within the pond area. The Waitangi Tribunal reported in February 1978, finding the claims of a Treaty breach well founded. It was the first report produced by the Waitangi Tribunal on an environmental matter.

By this time, the Electricity Department had abandoned its plans for the power station. The proposition simply didn't stack up, particularly in the face of such strong opposition. The *Auckland Star* summarised the position on 5 July 1977 in an article headlined 'Waiau Pa farce'. It stated, 'the station, it may be recalled, was intended to cater for a growth in demand that hasn't occurred, by burning up our valuable Maui gas as rapidly and wastefully as possible'.

As Roger Wilson observed, 'thus the battle over Auckland Thermal No 1 was won — and it can be counted a true environmental victory, for if there had been no opposition, there is little doubt that it would have been built'.

In the June 1977 edition of *EDS News,* Mike O'Sullivan wrote, 'many people, even sympathisers, told us in 1975 that "Auckland Thermal No 1" could not be stopped. We believe this important battle has proved that if local people organise to use the political and legal systems effectively they can defeat bureaucratic aggression, to the benefit of both their own region and the nation.'

When the government passed new town

The intertidal area adjacent to Clarks Beach and Waiau Pā remains undeveloped due to the efforts of EDS, the Whātāpaka Marae and the Manukau Harbour Protection Society. RAEWYN PEART

and country planning legislation in late 1977, it quietly removed the requirement in the Clean Air Act that bound the Crown to comply with planning law, thereby effectively legislating to overturn EDS's Supreme Court victory. Sir Guy Powles made strong objections on behalf of EDS, both against the substance of the change and the clandestine way in which it was made. EDS claimed that 'the proposal is purely and simply designed to free the Crown from any possible legal constraints on its executive power'. But the Crown went ahead anyway.

At a gathering on Whātāpaka Marae in early January 1978, chairman of the Manukau Harbour Protection Society, Noeline McLarin, and elder of the Whātāpaka Marae, T. E. Kirkwood, presented EDS with a cheque for $750. The money was sourced from funds raised at a gala day held at the marae and other community fundraising activities. The gift was an expression of gratitude to EDS for the support it had given their communities throughout the struggle to stop Auckland Thermal No. 1.

Reflecting back on the controversy, Mike O'Sullivan observed,

> I spent a lot of time going to public meetings and putting submissions in. It was great fun and very interesting as New Zealand was a pretty naïve society in those days. A group of people who knew what they were doing had an enormous influence. We could get access to politicians and news media. You could ring up the *New Zealand Herald* or TV reporters and arrange to give interviews. I remember going around to see Bill Birch, who was Minister of Energy at the time, at his home with one of the local Waiau Pā protest leaders. It was Sunday morning and Bill came to the door with a towel around his waist to face the wrath, having just come out of the shower. We felt quite sorry for him.

DURING THE 1980s EDS became heavily involved in another energy matter: the storage of liquefied petroleum gas (LPG) and associated risks to human life. The production of LPG was one of the multiple ways that government proposed to use the enormous reserves of gas discovered in the Kapuni and Maui gas fields off Taranaki. In persuading the private sector to develop the field, the government had agreed to a 'take-or-pay' clause in its joint-venture agreement, where it was committed to pay for a specified quantity of gas over a 30-year period irrespective of whether it was used or not. The government was therefore determined to use it.

The LPG, which was being produced in New Plymouth by Shell BP and Todd, was to be shipped by coastal tanker to a marine terminal in the Manukau Harbour. It would then be piped to a nearby regional storage facility awaiting distribution over the upper North Island. Liquigas Limited (a consortium of Shell BP, Todd and Rockgas) was the distributer of LPG which was then on-sold to retailers.

LPG is created by compressing gas to several hundred times its normal density. This makes it denser than air, meaning that LPG can travel as a heavy cloud over long distances, either along the ground or through underground passages and sewers.

Even if only a small fraction of LPG is released into the air it forms a flammable mixture which is potentially explosive. For example, in France in 1966, a frozen drain valve caused a leakage from an LPG tank located at a petroleum refinery in the suburbs of Lyon. The leaked LPG dispersed to a nearby road and was ignited by the engine heat of a passing car. This caused a fireball which killed 18 firemen and spectators and injured an additional 31 people. The fireball also caused the explosion of the LPG tank itself which, in turn, ignited other fuels on site. It took 48 hours to bring the conflagration under control. By that stage, the entire refinery was destroyed.

With such incidents in mind, EDS was concerned about the risks of storing large quantities of LPG in urban areas. The first site where Liquigas sought to establish the Auckland regional LPG storage facility was on the shores of the Manukau Harbour at Onehunga. But this met with such strong local opposition that the company withdrew its plans.

A second site was then identified at Ash Road in Wiri, which was some 2 kilometres away from the Manukau City centre. But it was much closer to other workplaces, with some

250 people employed within 300 metres of the proposed location, a figure anticipated to increase substantially in the future. Much of the area surrounding the Ash Road site was zoned for industrial use but had yet to be developed. It was expected that some 7200 people would be working within 1 kilometre of the LPG storage tanks once all the properties had been built on.

In late 1981, Liquigas applied to the Manukau City Council for consent to establish the storage facility at Ash Road to store up to 3000 tonnes of LPG. EDS and others objected, and the council turned down the application due to safety concerns. Liquigas then appealed to the Planning Tribunal, a specialist national judicial appeals body established under planning legislation. The appeal was heard in March 1983 and lawyer Richard Brabant appeared for EDS. The main focus of the hearing was

The bulk LPG terminal which was established on McLaughlins Road, Wiri, Auckland, has adopted rigorous safety measures as a result of EDS's efforts. In particular, the storage tanks have been buried and are covered in a rock mound, as can be seen in this image. RAEWYN PEART

the potential impact of the proposed LPG storage facility on the safety of people in the area. Much technical evidence on risk, and the potential catastrophic impacts of an accident, was presented by both sides.

Evan Penny, who subsequently become a director of EDS, was one of the witnesses who gave evidence for the society. Penny had been a senior Shell engineer, and while working for the company he had been involved in the selection of the Ash Road site. He had also undertaken the preliminary design of the proposed depot. Penny had resigned from the company in October 1980, before the planning application was lodged.

As Penny explained,

> I thought the industry was trying to keep the lid on the longer range hazards of LPG and wanted to dampen public alarm that might be caused by any such debate. I believed they were concerned about the significant costs worldwide associated with increasing safety isolation of bulk LPG facilities. Bob Mann wasn't persuaded that the LPG industry in New Zealand was doing everything it should be doing and I agreed with this view. He called at our house in Mount Eden after I had resigned from Shell and suggested I have a look at providing evidence for EDS. I was in a unique position to do so.

In effect, Penny had turned whistleblower.

After hearing much competing evidence, the Planning Tribunal concluded that although the design of the depot met or exceeded all relevant regulations and codes of practice, it was not possible to entirely eliminate the risk of an unintended release of a substantial quantity of LPG. And if such a released occurred, the LPG would inevitably be ignited and create an explosion or very intense fire. This, the tribunal found, could endanger people and property up to 550 metres downwind, which at the proposed site could result in many deaths and even more injuries.

Despite this finding, the majority of the tribunal members granted consent. However, this was subject to conditions, one of which required Liquigas to include adjacent land (which it didn't own) within a buffer zone for the site. The chairman of the tribunal, Judge David Sheppard, dissented. He thought the risk was too great and that there were other potential sites which were more remote and would reduce the number of people exposed to danger. In his view, consent should be declined.

Former Shell engineer Evan Penny supported EDS's case to obtain more robust safety measures for the LPG storage facility at Wiri, Auckland. Penny was a director of EDS from 1983 to 1984 and later became an Environment Waikato councillor. RAEWYN PEART

The buffer condition effectively required Liquigas to buy or lease the adjacent site. However, as it later turned out, the company was unable to reach agreement with the adjoining landowner. This prompted the company to make a fresh application for a more remote site in Wiri on McLaughlins Road. This time the council approved it and the decision was upheld by the Planning Tribunal on appeal by EDS. It was at this point that Liquigas sought $86,000 costs against the society, a very large sum for a tiny not-for-profit organisation. Although not accepting

Richard Brabant was admitted to the bar in 1972 and he subsequently developed a law practice in town planning. In 1979, he was elected as a member of the Auckland Regional Authority and he served for two terms until 1985. In 1992, Brabant became a barrister specialising in environmental and local government law. He is now retired from legal practice.

RAEWYN PEART

Brabant was a director of EDS from 1983 to 1986. He ran many EDS cases including the LPG cases during the 1980s, the combined-cycle gas-fired power station cases in the early 2000s and the renewal of water rights for the Huntly Power Station. These were long-running cases and Brabant expended a considerable amount of time and energy on them on behalf of EDS for no payment. At times it was difficult to juggle his commitments to EDS with those of his private legal practice.

As Brabant recalls, 'we had no money so you had to make do with what you had. Often we didn't have the resources to engage witnesses, or couldn't find a witness to deal with a particular technical issue, so you had to rely on reading and understanding the technical evidence of the applicant and cross-examining on it. As you didn't have the help of your own witnesses you had to become knowledgeable on the topic yourself. It was incredibly good training for me. I ended up at a very young age appearing in cases with other lawyers twice my age. I had some pretty torrid times with [Judge] Sheppard but it was a great learning curve.

that such a large amount was warranted, the court did award $14,000 against EDS in favour of Liquigas. It was one of the few substantial costs awarded against the society. The manager of Liquigas was subsequently persuaded not to pursue payment.

On what the LPG cases achieved, Evan Penny observed, 'we forced the company to shift the depot away from sensitive uses to vacant land around the Wiri train depot. The company also agreed to mound the storage vessels which was a big win. And overall, the debate that EDS raised about safety issues caused a culture shift throughout the whole enterprise.'

There were also important developments in the law. As Richard Brabant explained, 'an interesting principal that came out of the LPG cases was that people in the workplace were entitled to equal protection from hazard as those living in their homes. It was a very important principle that EDS established then.'

3. Thinking bigger

BY THE LATE 1970s, EDS had gone through a transformation. Founder David Williams had resigned from the board in 1974 due to the pressure of work commitments. After moving its headquarters out of Russell McVeagh in 1973, EDS had set up a small office in donated space at the Auckland Law School, then in 1975 it moved to office space provided by the National Bank in Shortland Street and finally, in late 1976, to the third floor of the Auckland Town Hall.

IN ITS EARLY DAYS EDS's office facilities were rudimentary. In the November 1977 issue of *EDS News*, the society put out a call for office furniture, stating 'The EDS office is still badly short of filing facilities (cabinets, folders) and shelving. Even bricks and more boards would be a help if anyone has any to spare.' EDS's activities were still largely based on voluntary effort.

In 1978, 31-year-old Gary Taylor had taken on the full-time role of co-ordinator, providing the society with some dedicated impetus. He was paid $5 an hour. Taylor was neither a lawyer nor a scientist but he brought a new set of skills to the organisation. Graduating with a Masters degree in history from the University of Auckland, Taylor had briefly been a school teacher before turning his hand to journalism working for Radio New Zealand. Just prior to joining EDS, he had been editor of the *Forest Industries Review*.

Taylor had become an environmental advocate almost by accident. With his artist wife, Mary Taylor, he had bought a house in the Waitākere Ranges and grew to love the bush. He then learnt of a plan to build a rubbish tip in the beautiful Waitākere River Valley and was outraged.

Taylor was introduced to EDS through Dick Bellamy. He had first met Bellamy as a reporter. They subsequently worked together on a campaign mounted by the Waitākere Ranges Protection Society to purchase Lake Wainamu on Tāmaki Makaurau Auckland's west coast. Taylor helped edit a publication for the campaign and Mary Taylor provided some illustrations. As Taylor recalls, 'I decided at that point there were more important things than making money.' Bellamy remembers Taylor telling him that he wanted to work for EDS. 'I said I wouldn't recommend it, as there was little money and no steady income. But he was not dissuaded.'

Ian Cowper, who had been one of the first student directors of EDS when studying at Auckland Law School, and was now a law partner at Morpeth Gould, re-engaged with the society during this time. He worked closely with Taylor, helping out with the administration of EDS, as well as with legal matters. Together they re-injected new energy into the organisation.

By this time, New Zealand's economy was struggling. The oil crisis had worsened: oil-producing nations through OPEC raised oil prices and the 1979 Iranian Revolution caused the withdrawal of significant quantities of oil from the world market. On top of this, by the late 1970s, New Zealand was suffering from the loss of Great Britain as a major agricultural export market after it joined the European Economic Community.

It was in this context that Robert Muldoon's National government conceived of the 'Think Big' strategy whereby the government would support big energy-intensive projects to reduce reliance on imports and improve the country's balance of payments. One of the Think Big projects was the construction of the country's second aluminium smelter which was proposed for Aramoana (the first smelter at Tiwai Point near Bluff having opened in 1971).

—ooo—

THE PROPOSAL to build a smelter at Aramoana was made public just two weeks before the 1972 election. It was floated as a vote catcher for National, but the party was defeated at the

Previous pages: The Clyde Dam on the Clutha River. This hydro power project was the subject of widespread public protest and a massive effort by the EDS legal team. ROB SUISTED/NATURESPIC

Ian Cowper (left) and Gary Taylor (right) presenting on behalf of EDS at the Parliamentary Select Committee on the National Development Bill in 1979. Their submissions achieved some very important amendments to the Bill which enabled EDS to subsequently challenge the Bill's application to the Aramoana Smelter and the Clyde high dam.
EDS

polls and the project ground to a halt. The idea reappeared two years later and was expanded to include an adjacent zinc smelter and car assembly plant. But the project was rejected by the Labour government at the end of 1974 on the basis of insufficient energy to supply the smelter. The government noted that as the proposal 'imported raw materials with the end produce allocated for export ... we would simply be exporting (hydro-electric) resources at a very low price'.

By the late 1970s, National was back in power and was proposing a National Development Bill as a way of ensuring that projects that met with government approval were fast-tracked through the planning process. EDS made detailed submissions on the Bill. Gary Taylor and Ian Cowper flew to Wellington in mid-November 1979 and presented the EDS submissions to the Lands and Agricultural Select Committee. They argued that the Bill was unnecessary, and that a better approach would be to fix up existing planning procedures.

EDS was concerned that the proposals concentrated unwarranted power in the hands of the Minister of National Development and reflected 'a disturbing shift' towards government by executive decree. If the Bill was to be retained, EDS argued that it should include a right of review of planning decisions by the courts, a redefinition of what was meant by the term 'national development work', and a strengthening of the role of the Commissioner for the Environment.

When the Bill was reported back to the House in early December, EDS and others had achieved some significant amendments. It now explicitly excluded application to nuclear energy, the fast-track provisions of

Aerial view of the outer Otago Harbour in 1962 showing Taiaroa Head in the foreground, which was and still is the country's only mainland royal albatross colony, and the rich Aramoana salt marshes and inter-tidal area near the centre of the image. It was these low-lying areas that were to be developed for the proposed aluminium smelter. WHITES AVIATION, ALEXANDER TURNBULL LIBRARY

the Act could only be applied to a work if it was 'essential a decision be made promptly' and, of critical importance to EDS's ability to later challenge projects coming within the ambit of the legislation, decisions could be judicially reviewed in the Court of Appeal.

The Act became law in mid-December 1979, at the same time as the proposal to build a second aluminium smelter re-emerged. The smelter proposition was based on an expectation that there would be surplus electricity-generating capacity over the subsequent 15 years which could be made available to energy-intensive industries in the South Island. However, it was soon evident that inaccuracies in government energy forecasting, which had been highlighted by EDS some years previously, had still not been addressed.

When the first government Energy Plan was released in 1980, it found the surplus to be much smaller than earlier thought, meaning that a vast programme of power station construction would be necessary to supply the smelter. But by this time the government had become wedded to the idea of a second smelter, and it forged ahead with the proposal, notwithstanding the energy implications.

By July 1980, the government had signed a memorandum of intent with a consortium formed between the New Zealand construction company Fletchers and the Swiss aluminium company Alusuisse. In December, the consortium (now formally incorporated as South Pacific Aluminium Limited) announced that Aramoana would be the site for the new smelter. Concerned locals formed the Save Aramoana Campaign to fight the plan. They developed innovative ways to raise funds, including declaring Aramoana to be an independent state and selling Aramoana passports, flags and other souvenirs at a 'border post'. This was followed by a travelling 'embassy' that roamed the country, raising awareness and selling citizenship certificates and postage stamps.

On the face of it, Aramoana was an attractive location for an aluminium smelter, particularly as the Otago Harbour Board owned an area of flat land immediately adjacent to deep, sheltered water suitable for berthing ships. But this part of Otago Harbour, which was near the harbour mouth, was also very important for wildlife. It had one of the very few remaining large salt marshes in the South Island, with its rich tidal flats supporting large populations of worms and shellfish, which in turn fed seabirds. Just across the harbour, at Taiaroa Head, was the country's only mainland royal albatross colony. There were also some 125 houses and holiday homes at Te Ngaru and Aramoana that would need to be removed to make way for the smelter.

Things started moving quickly and in April 1981 an Order in Council was issued to bring the project under the provisions of the newly minted National Development Act (1979). An application for the smelter was quickly submitted to the Planning Tribunal, which was tasked with conducting an inquiry under the Act. However, the tribunal's findings were only to be recommendatory, with the final decision resting with the Governor-General in Council, which was in effect government ministers.

It was at this point that EDS stepped into the fray. Both EDS and Forest and Bird were concerned about the environmental damage that would be wrought on the fragile coastal environment at Aramoana, where 25 hectares of biologically rich intertidal flats were proposed to be reclaimed. But there was also the broader energy picture to consider. The enormous power needs of the smelter would necessitate construction of additional power stations and the environmental groups feared that this would drive the damming of multiple rivers and also, potentially, the establishment of nuclear power plants. Gary Taylor firmly believed that it was 'in the national interest to stop the project from going ahead'.

The acclimatisation societies also recognised the potential threat to the nation's rivers from a second aluminium smelter and mounted strong opposition to the government's plans. This included a national advertising campaign. One particular advertisement featured a photograph of the Minister for National Development Bill Birch accompanied by the words 'To buy his smelter will we have to sell our country?'.

In 1980, Forest and Bird had agreed to provide EDS with badly needed financial support in exchange for legal assistance, and so EDS represented both organisations in most of the legal proceedings associated with the smelter.

EDS filed the first proceedings in 1981. These challenged the decision to apply the provisions of the National Development Act to the smelter project. The legal challenge was on a number of grounds, including that there was no need to put the smelter on the fast track as it was not 'essential' that a decision be made quickly, as required by the Act, and that the procedures undertaken to determine the Order in Council had been faulty. Similar proceedings were lodged by the Coalition for Rational Economic and Environmental Development in New Zealand (CREEDNZ) Inc, which had been formed to pursue legal proceedings against the smelter, and Garry Holden of Te Ngaru.

The first matter heard by the Court of Appeal was an application by EDS and Forest and Bird for discovery of cabinet documents

The beach at Aramoana 2008. The proposed aluminium smelter would not only have destroyed important ecological areas but some 125 houses and holiday homes at Te Ngaru and Aramoana were to be removed to make way for the smelter. RAEWYN PEART

that informed the decision to issue the Order in Council. This was a radical move. At that time, cabinet documents were extremely secret and kept well away from the public eye. The environmental groups were represented by former law academic and barrister Dr George Barton QC, as well as R. A. McGechan, who later went on to become a High Court Judge. The Solicitor-General vigorously opposed the application on behalf of the Minister of National Development and the Governor-General.

This was the first time that a New Zealand court had been asked to decide a question relating to Cabinet or Executive Council papers. Consequently, there were no legal precedents within the country to provide guidance and the Court of Appeal had to turn to decisions of the House of Lords in England and the High Court of Australia.

After carefully considering the matter, the Court concluded that it did have jurisdiction to order the documents to be handed over, and that they should be provided to the judges for scrutiny. This was on the basis that

> If parties such as the present plaintiffs were denied all access to the respondents' documents it could in practice be virtually impossible to challenge an Order in Council under the National Development Act on any grounds going to the reasons for the Order. ... These cases are of major public importance. Public confidence in the administration of the Act and in judicial safeguards would be shaken if the Court were to confine the scope of review so narrowly as to invite suggestions of rubber stamping.

The decision sent shock waves through Wellington. It was the first time the courts had opened the doors of Cabinet and ordered the production of documents. As former EDS director and Court of Appeal Judge Tony Randerson recalls,

> the smelter case was extraordinary. Apart from the environmental issues involved with the smelter, there was a big stand-off between the Court of Appeal and the Crown over whether Cabinet minutes should be disclosed to the court. In the end the court insisted the government produce the minutes which they claimed were confidential. It was a constitutional moment as to who was going to win: Cabinet or the courts. But the courts won and got the papers.

Although EDS and Forest and Bird subsequently lost the case on its merits, they had sent a strong message to government that it needed to open itself up to greater public scrutiny. The following year, the Official Information Act 1982 came into force and for the first time created a legal presumption that official (government) 'information shall be made available unless there is good reason for withholding it'. Cabinet papers are now routinely proactively released.

After the Court of Appeal proceedings had been disposed of, the planning application for the smelter continued down the National Development Act track. An environmental impact report was produced and scrutinised by the Commission for the Environment. EDS was soon back in the Court of Appeal challenging the validity of the report on the basis that it did not address the wider implications of electricity consumption by the smelter, and, in particular, the extent to which it could incentivise further hydro development on wild and scenic rivers. The society's case was argued by then-prominent barrister and later High Court Judge Peter Salmon.

The proceedings again failed, but the court made useful comments on the meaning and importance of environmental impact reports, including that they should include 'adequate and reliable reference to every matter that is significant and relevant' and should not 'avoid awkward or significant environmental issues'. EDS escaped an order of costs on the basis that the proceedings had 'been brought by the Society in the hope that the issue raised is in the public interest'.

Although EDS had not been successful in opposing the smelter in the higher courts, progress had still been made. As explained in the December 1981 issue of *EDS News*,

> not the least achievement in mounting these cases is surely to show Government and large organisations that members of the public can mount serious challenges to activities of both Government and private industry and that the Courts and

> indeed the public will carefully scrutinise the actions of the proponents of major developments. While we may have lost the individual battles it certainly seems as though significant advances have been made in the war.

Once the proceedings in the higher courts had been resolved, the Planning Tribunal hearing on the merits was set down for early March 1982. But when the day of the hearing arrived, South Pacific Aluminium announced that it would not be proceeding with the application. Alusuisse had withdrawn from the consortium, ostensibly because the power price demanded by the government was no longer attractive due to the rapid softening of the world aluminium market. It was not clear how much the strong opposition to the project, and multiple legal challenges by EDS and others, contributed to the decision. Although the government tried to find a new partner for the project, it was not successful and the proposal was never revived.

The smelter project created the first links between EDS and the acclimatisation societies (subsequently Fish and Game Councils) which have continued up until today. As former Fish and Game national chief executive Bryce Johnson observed,

> The fact that EDS was there, and very vocal in the public arena, helped the acclimatisation societies survive. It was the first time we had really confronted the government of the day and it would have been an easy thing for the government to legislate us out of existence. At that time, there was really no-one else out there other than the two of us. It was the beginning of an association between the two organisations of mutual respect.

The land at Aramoana was subsequently passed on to the Department of Conservation and declared an ecological area of national importance due to its 'impressive vegetation sequence spanning mud flats, saltmarsh, salt meadow and salt-tolerant shrubland – the most extensive ecosystem of its kind left in Otago'.

The link that EDS and others had drawn between the Aramoana smelter project and the prospect of damming rivers was a real one, as the government used the Aramoana smelter as the primary justification for constructing a high dam on Mata-Au Clutha River at Clyde.

THERE IS A LONG HISTORY to plans to develop a series of dams on the Clutha River. The Clutha is the longest waterway in the South Island, draining nearly one-twelfth of the country's land surface. It is fed by three glacial lakes on the edge of the Southern Alps – Wakatipu, Wānaka and Hāwea – before flowing through the Otago high country and down onto the plains. The river finally enters the Pacific Ocean at Balclutha, south of Ōtepoti Dunedin.

Investigations into options for using the river to generate hydro-electric power began in 1962, with up to 16 dam sites studied. By 1974, two clear options had emerged, schemes F and H. Scheme H was favoured by the Clutha Valley Development Commission, which had been established to investigate the effects of hydro-electric power development on the resources of the Clutha Valley, and which reported to the Commissioner of Works in November 1974.

Scheme H was also favoured by the Clutha Valley Advisory Committee, which was set up to re-examine the options in October 1976. Schemes F and H both contemplated a series of dams, and flooded a similar area of land,

This aerial view of Otago Harbour highlights the high natural values of Aramoana wetlands, (lower right) which has now been declared an ecological area of national importance under the management of the Department of Conservation. It would have looked very different if the government had succeeded in its bid to build an aluminium smelter there.
LLOYD HOMER, GNS SCIENCE

but their effects on the Cromwell Gorge were starkly different.

The Cromwell Gorge was renowned for its scenic beauty and productive apricot orchards. Scheme H included a low dam below the historic mining town of Cromwell which sat upstream of the Cromwell Gorge, and another dam above the smaller village of Clyde, downstream of the gorge. Scheme F replaced these two dams with one, much higher, dam just above Clyde. This would raise the water level of the hydro lake by a further 30 metres and flood a much larger area of the gorge, including all 86 hectares of lush apricot orchards, the railway and road, and parts of Cromwell itself.

In 1975, the Labour government had announced that Scheme H would proceed. But after the National Party won the general election later that year, it opted for Scheme F, which included the high dam at Clyde. The ostensible reason was that the high dam was expected to be $29 million cheaper than constructing the two lower dams. It was also scheduled to provide full power output four years earlier than Scheme H, and, importantly, in time for the anticipated energy needs of the then prospective aluminium smelter at Aramoana.

There were also other benefits anticipated by the government from the earlier completion of Scheme F. It would release construction workers sooner so they could be deployed onto other hydro proposals. As the subsequent Ministerial Review Committee into the dam project decision-making process observed, 'Nothing could better demonstrate the climate of the

The Cromwell Gorge, shown here in 1947, which was proposed to be flooded by the high dam in Scheme F of the hydro-electric power development proposed for the area. The image shows the road, steep rocky walls and extensive orchards on the upraised river flats that were all to be lost under the hydro lake.
VC BROWNE & SON

times. New dams in endless vistas were envisaged stretching out ahead of MWD [Ministry of Works and Development] and the Government into the infinite and indefinite future.'

The Clyde Dam was to be the biggest concrete dam in the country's history. As Prime Minister Robert Muldoon later stated in Parliament, 'There is the Clutha running down to the sea. Let us put a dam across it and make some electricity that we can sell as energy.' Before the dam could proceed, the government needed water rights. But there was concerted local opposition with a Clutha Action Committee being formed and Clutha Rescue subsequently occupying the banks of the Clutha River near Clyde. In addition, Cromwell Gorge orchardists were not about to give up their land without a fight.

At the time the Clyde Dam controversy erupted, EDS director Stephen Mills had recently returned from the United States to a position at the Auckland Law School. He had just completed a Masters in Law degree at the University of Pennsylvania and a year as teaching instructor at the University of Michigan Law School. While there, he had contact with Joseph Sax. Sax was one of the new breed of lawyers who focused exclusively on the environment, and he was credited with helping to shape environmental law in the United States. Sax proposed the idea of a 'public trust' with citizens having the right to sue government, business and private individuals to protect the natural environment. It was inspiring stuff.

Concerned about the Clyde Dam proposal, Mills undertook a forensic examination of the wording of the Water and Soil Conservation

Act, after which he concluded that it might be possible to mount a successful legal challenge to the Clyde Dam water rights. He raised the issue with then EDS executive director Ian Cowper and it was discussed by the EDS Board. Mills recalls,

> Ian's response was that if we were to win, the government would just change the law. But I said 'wait a minute, EDS was set up modelled on EDF and our role was to test the law in important areas. And if the government changes the law afterwards, well that's what they do, but it's no reason for not challenging the legality of the action.' Ian replied, 'well alright then, you can go down there'. If we could get local support, EDS was on for it.
>
> I had two little boys at the time. Shonagh Kenderdine made her crib available in the middle of nowhere and I took the family down for the summer. I spent my time tootling up and down the Clutha, talking to orchardists whose land was to be flooded, to see if we could get their support. I then needed to find someone to run the argument as I didn't have the necessary level of court experience. Chris McVeigh a Christchurch barrister, and subsequently a QC, said he would do it.

In June 1977, the Minister of Energy lodged an application for water rights for the high dam. Under the normal water rights process applications were to be lodged directly with the regional water board for decision. But the statute provided a different process for Crown water rights. These applications were lodged directly with the Minister of Works and Development who, in turn, referred them to the National Water and Soil Conservation Authority for a decision. The Authority was chaired by the Minister of Works and Development, the same Minister who received the application. For Crown water rights, the regional water board was given the opportunity to make recommendations but was stripped of decision-making power. This meant that the government was, in effect, applying to itself for water rights. It was far from an even playing field.

In both cases, decisions could be appealed to the Planning Tribunal, but for Crown water rights the standing provisions were much more restrictive. Only those non-public bodies who were 'detrimentally affected' by the decision could appeal. The public interest which EDS represented, was given standing in appeals on normal water rights but not for those sought by the Crown. This further weighted the process in favour of the government. It also meant that, although EDS could be heard at the lower-level hearing, it could not lodge appeal proceedings in its own name.

The water right application for the Clyde Dam was first sent to the Otago Regional Water Board for consideration. After receiving 209 submissions (of which only three were not opposed) and conducting an informal hearing, the water board recommended that the water rights for the high dam be declined, and those for the two low dams (corresponding to Scheme H) be granted. But this did not accord with the government's plans. The National Water and Soil Conservation Authority overturned the regional water board's recommendations and, in December 1977, granted the government a water right for the high dam. Backing the owners of land which was to be flooded, EDS lodged appeals with the Planning Tribunal, both on their behalf and in EDS's name.

At this point, EDS also lodged judicial review

proceedings challenging the decision of the National Water and Soil Conservation Authority.

This challenge was on a range of grounds, including failing to pay proper regard to the recommendations of the Otago Regional Water Board, failing to conform to the intent of the Water and Soil Conservation Act, taking in account extraneous matters and bias. But when the court issued a decision in early October 1979 it concluded that the grounds had not been made out and the legal action was dismissed.

While the appeals were waiting to be determined by the Planning Tribunal, construction of the dam had gone ahead, despite no water rights being confirmed. In fact, work at the site had commenced in 1976, prior to the water right application even being lodged. This served to underscore the arrogant stance of the government when it came to the law. The impacts of not proceeding with the dam on 'the very large workforce that has been assembled on the site during the past two years', was to later be used by Energy Minister Bill Birch as one of the 'compelling reasons' for passing special legislation to authorise the high dam in 1982.

EDS requested that the government halt work on the dam, pending the hearing of the Planning Tribunal appeals, but it declined to do so. EDS then mounted legal proceedings in the High Court, this time in an attempt to stop the works. Unfortunately this also failed, after the judge accepted the government's claim that the works would be equally relevant to the low dam option if water rights for the high dam were declined.

The Planning Tribunal hearing on the water rights finally commenced in September 1980, following delays requested by the Crown. The tribunal had six members and, unusually, two were judges — Judge John Treadwell, who was the chair, and Judge Peter Skelton. In the end, some of the issues were so contested that the Planning Tribunal members could not reach a unanimous decision — an unusual event. The tribunal's split decision was released on 16 December 1980, with a majority granting the water rights.

At the commencement of the hearing, the standing of EDS to lodge an appeal was successfully challenged and the society's appeal was dismissed. However, anticipating such an outcome, EDS had also lodged appeals in the names of the owners of land that was proposed to be flooded, as they were clearly affected by the dam and therefore had standing.

One of the early matters dealt with by the tribunal was the end-use of the power. The tribunal noted that the original impetus for the government's adoption of the high dam, and the approval of it by the National Water and Soil Conservation Authority, was projected electricity demand. But this was subsequently found to be over-estimated by almost 100 per cent and could no longer be used as justification.

The government belatedly identified the aluminium smelter at Aramoana as the end-user of the power. In hindsight, the Crown's requests to delay the Planning Tribunal hearing may well have been motivated by the need to secure investors in the Aramoana smelter project in order to shore up the Crown's case on the urgent need for the high dam.

The Planning Tribunal concluded 'on present power forecasts that the necessity for DG3 (high dam), with its 1988 commissioning date for the first machine, is solely caused by the requirement to provide power for a new smelter'. However, the appeals were to be determined within the ambit of the Water and Soil Conservation Act, and the tribunal found that this did not permit consideration of the end-use of the power generated.

When it came to the inundation of land,

EDS lawyer Chris McVeigh argued on behalf of the landowners that the low dam was to be preferred, as it avoided flooding of highly productive fruit-growing land. Twenty per cent of the national production of apricots came from the Cromwell Gorge, with an annual production value of more than $1 million. The tribunal accepted that the orchard land was 'premier land of the highest quality' with a micro-climate that made it virtually frost free. This climate enabled disease-free fruit to be grown, and they ripened 10–14 days before other apricot crops in the country, giving the Cromwell orchards a profit advantage.

When weighing up the losses from the dam proposal, the tribunal members could not agree. A majority of four held that the granting of the water rights for the high dam outweighed the resultant loss of horticultural and other land and associated scenic and environmental values. But the two judges on the tribunal dissented. Judge Treadwell considered that the hydro power required could be obtained without flooding the orchards through the construction of the lower dams, and therefore 'the land which is of high actual and potential value for the production of food should be saved'. He was opposed to granting the water right.

Judge Skelton noted that, despite the water rights not yet being granted, government had already spent some $45 million on the scheme, including building a new bridge at Cromwell, rebuilding and realigning the part of the state highway that would be inundated by the high dam and undertaking considerable work at the dam site itself.

However, the judge held that such expenditure should not influence the tribunal's decision. Skelton went on to analyse the difference in power output between the high and low dam options. Noting that the increased power output of the high dam was just 4–5 per cent, he found that such a small increase in the nation's power supply did not justify the loss of the productive horticultural and agricultural land. For this reason, he also opposed granting the water right.

On reading the decision, one gets the strong impression that the two judges were distinctly unimpressed by the government's assumption that the grant of water rights was a foregone conclusion. This was evident by the work to build the high dam commencing prior to obtaining the requisite consents.

EDS (under the orchardists' names) again appealed to the High Court, this time on the basis that the tribunal had failed to consider the end-use of power. This was the legal challenge that Stephen Mills had identified at the outset.

In a judgement issued on 13 May 1982, Justice Casey upheld the appeal. This was on the basis that there was nothing in the Water and Soil Conservation Act that precluded consideration of the power end-use and in this case it was highly relevant because 'the future existence of the Aramoana smelter is at the very heart of the decision favouring a high dam, with its widespread inundation'. The High Court set aside the tribunal's decision and told it to have another look at the situation. This was a major win.

The Planning Tribunal hearing resumed on 2 August 1982. By this time Alusuisse had withdrawn from the smelter consortium and the application for planning consent under the National Development Act had been withdrawn. The government's sole end-user of power from the Clyde high dam had now gone. The tribunal ruled shortly afterwards that a water right would not be granted. This time, the six tribunal members were unanimous. In the absence of a large user, the government had failed to demonstrate that the power was needed.

Cromwell in Central Otago during the 1980s, prior to being flooded by the Clyde high dam. The government justified passing special legislation to overturn adverse court decisions on the basis that Cromwell had become dependent on the construction of the dam.
ERIC W. YOUNG, AUCKLAND LIBRARIES

By this time EDS and others had been involved in five separate legal hearings, three in front of the High Court and two in front of the Planning Tribunal. It had been a massive effort by the EDS legal team. A subsequent Ministerial Review Committee observed that 'there is probably no project in energy history in New Zealand which has been more controversial for longer than the high dam at Clyde'. EDS had used every legal tool available, and in the end had won. Or had it?

—OOO—

AFTER THE WATER RIGHTS were turned down, Energy Minister Bill Birch stated that the tribunal was not equipped to rule on matters that were really government policy. The government, which by now was very committed to the high dam option, confirmed that it would pass special legislation to authorise the high dam. It later turned out that cabinet had been discussing special legislation to overcome any unfavourable legal proceedings since at least March 1978. The government appeared to have had no intention of abiding by the courts' decisions.

On 24 August, a week after the Planning Tribunal issued its decision, Minister Birch introduced the Clutha Development (Clyde Dam) Empowering Bill into the House. He admitted that around 600 men were already working on the dam site, and that the government had already spent between $80 and $100 million dollars on the dam, despite not having the requisite water rights to authorise its operation. Cromwell had become dependent on construction of the dam, with any interruption likely putting hundreds out of work.

But the government was on shaky ground. After the 1981 general election, which focused on the 1981 Springbok Tour, National lost considerable electoral support and was left with a wafer-thin majority of one. Two national MPs, Marilyn Waring and Mike Minogue, refused to support the Clyde Dam empowering legislation which meant that National did not have a majority to pass it. In Mike Minogue's words, the government was 'getting to the stage where it doesn't regard itself as bound by the due process of law at all'.

The Labour Party also opposed the legislation, highlighting that the government had no end-use for the power. Social Credit leader Bruce Beetham had been vocal in the party's opposition to the special legislation stating that it 'was close to treating Parliament and the courts with contempt'.

On the evening that the Bill was having its third reading, Gary Taylor was in Wellington and he decided to watch events in Parliament in person. He had earlier spent time lobbying Bruce Beetham and was expecting the Bill to be voted down. But to his and many others' surprise, the two Social Credit MPs — Bruce Beetham and Gary Knapp — voted in support of the Bill. It later became apparent that they had reached a deal with the National Party, voting for the Bill in exchange for National's support for Social Credit policies, including irrigation in Central Otago. The Act came into force on 30 September 1982.

Clyde Dam protest sign on the padlocked doors of the Court of Appeal in Wellington in October 1982. After the government legislated to overturn adverse High Court and Planning Tribunal decisions in order to proceed with the Clyde high dam, protesters padlocked the doors of courthouses and put up 'Dam Democracy' notices.
DOMINION POST 1 OCTOBER 1982. PHOTOGRAPHIC NEGATIVES AND PRINTS OF THE *EVENING POST* AND *DOMINION* NEWSPAPERS. REF: EP/1982/3371-S-F. ALEXANDER TURNBULL LIBRARY

The deal did considerable harm to Social Credit's popularity, as most of the party's supporters opposed the Muldoon government, and the Clutha high dam proposal in particular. Social Credit had secured 20 per cent of the votes in the 1981 general election, but this dropped to less than 8 per cent in 1984. At that election, Social Credit leader Bruce Beetham lost his seat in Parliament and a year later the name 'Social Credit' was dropped altogether from the political landscape.

In supporting the use of special legislation to overturn the courts, Otago National MP Warren Cooper wrote 'Finally, who runs New Zealand? Are we to abrogate the whole of decision-making process to the courts ... Legislation to give approval draconian, but oh how necessary!'

It was the abrogation of the court process that outraged many New Zealanders. The passage of the Bill resulted in an outpouring of fury. Protesters wrapped chains around the handles of the Dunedin Courthouse and attached a sign stating 'This Court is now obsolete, irrelevant, and just a nuisance. Accordingly it is CLOSED until such time as people no longer expect the law to protect their rights.' Identical 'Dam Democracy' notices were affixed to the padlocked doors of the Court of Appeal in Wellington and the Christchurch High Court.

Gary Taylor summed up the situation in the EDS 1982 annual report.

> For Parliament to legislate to override the general law of the land and, more particularly, the fruits of litigation by affected landowners, is a very grave step indeed. It is constitutionally unacceptable and undermines the rule of law. How can affected parties oppose ill-considered projects in the future through the judicial process with any confidence? It encourages extra-legal approaches to opposition and is a matter that Government should reflect upon. You cannot give people rights and then take them away because you don't like the result.

The Clyde Dam is New Zealand's biggest concrete dam. Its history is riven with controversy due to the government's determination to go ahead with the dam despite the fact that it lacked the requisite water rights to authorise its operation. EDS and others were involved in five separate legal hearings to oppose the dam and the water rights were eventually declined. The government then legislated to overturn the courts with support from Social Credit, a move that spelled the end of Social Credit as a political party. ROB SUISTED/NATURESPIC

The Clyde Dam under construction. There were multiple problems during the construction phase that almost doubled the cost of the dam and delayed completion by five years. This was largely due to the lack of proper scrutiny by engineers of the high dam option before it was adopted by the government. LLOYD HOMER, GNS SCIENCE

Construction of the dam proceeded, but it was not smooth sailing, despite legal obstacles having been overcome. The assumption that the Scheme F high dam would be the cheap and fast option was proven to be patently false.

The original contract for construction of the dam provided for the excavation of 14,000 cubic metres of rock to provide a solid foundation for the dam wall. But when an extensive area of weak rock was discovered, a staggering twenty times that amount had to be eventually dug out.

Then there were the active fault lines close to the site. Although they were identified by geologists during the late 1970s, this information was not effectively communicated to the dam design engineers. It was only when a US engineering geologist visited the site in early 1982, after construction was well underway, that the dam engineers were alerted to possible earthquake risks. They responded by redesigning the dam, which decreased its generating capacity by a quarter, and put it below the generating capacity of Scheme H, the two-dam option.

Although there were known landslides in the Cromwell Gorge, they were thought to be inactive. But during construction it was found that some were slowly moving downhill towards the reservoir area. Managing the landslide risk became so complex, that up to 40 geologists spent two years identifying and stabilising landslides before the filling of Lake Dunstan could commence in 1992. Stabilisation required a complex process of drilling tunnels

RAEWYN PEART

Stephen Mills KC completed a Bachelor of Laws (Hons) from the University of Auckland before working at law firm Russell McVeagh for several years. He developed a particular interest in environmental law, so headed to the University of Pennsylvania to complete a Masters, where he was awarded the American Trial Lawyers Association Prize in Environmental Law. On returning to New Zealand in 1974 Mills took up an academic position at the University of Auckland Law School. He was David Williams's successor in teaching the environmental law course. Mills was elected to the Devonport Borough Council for two terms and he played a key role in the steps the council took to block the proposed filling in of Ngataringa Bay for a marina housing development.

In late 1978, Mills took up an offer to teach environmental law at Lewis and Clark College at the Northwestern School of Law in Portland, Oregon. He has also taught at the universities of Western Ontario and Dalhousie. On returning from overseas, Mills became a senior litigation partner at Chapman Tripp and later become chairman of the firm. In 2002 he joined the independent bar and was appointed a Queen's Counsel in 2007. Mills specialises in contract, commercial, public and defamation law.

Mills became an EDS director in 1974 and served on the board until 1978. He developed the legal arguments that demonstrated that the Auckland Thermal No. 1 power station needed town planning consent and that won the Clyde high dam case for EDS and the landowners the society was representing.

into the shingle in order to dewater and stabilise the shattered rock.

In the end, the dam cost 44 per cent more than the original estimate and only started generating power in 1993, five years later than planned. In 1990, the Labour government commissioned a review of the Clyde Dam project. It found that there had not been a proper investigation process, with Scheme H having been selected for political reasons, despite it having been investigated less than the other sites which had been preferred by the engineers.

The Electricity Department told the review panel that 'beyond any doubt' the Clyde Dam would not have proceeded if the full costs had been known. But that was little comfort to the landowners who had lost their orchards and New Zealanders who could no longer experience the once stunning Cromwell Gorge.

Reflecting on the case, Stephen Mills observed,

> in the end the High Court did declare the Tribunal's decision unlawful and indeed Muldoon did move to pass special legislation. It was passed with votes from Social Credit which effectively destroyed that party. So EDS was responsible for getting rid of Social Credit and the high dam went ahead. It was a very big case for EDS and, yes, Ian was right. What he predicted did happen. But I still say it was absolutely the right thing to do.

4. Wild and scenic

THE CONCEPT OF PROTECTING RIVERS in their natural free-flowing state first arose in the United States, where concerns about water pollution and the construction of dams provided the impetus to pass the Wild and Scenic Rivers Act in 1968. Similar concerns arose in New Zealand during the 1970s, when the government undertook a vigorous dam-building programme and provided councils with financial support to build small hydro schemes around the country.

THE MINISTRY OF ENERGY had identified hydro potential on a number of highly valued recreational and wilderness rivers, such as the Mōtū, Whanganui, Mohaka in the North Island and the Kawatiri Buller, Rakaia, Māwheranui Grey, Waiau Toa Clarence and Whakatipu Kā Tuka Hollyford in the South Island. They all seemed to be at risk. As Gary Taylor recalled,

> this was part of the ongoing Ministry of Works juggernaut of building more hydro power stations in the country irrespective of whether there was a need for them or not. It was the time of 'Think Big' and the expectation was that there would be a number of big developments which would require energy, so they were on a bit of a roll.

In addition to the large Upper Waitaki and Clutha hydro schemes, the Ministry of Energy had a strong focus on developing North Island rivers. This was because the efficiency of North Island thermal and geothermal plants could be maximised if hydro plants serviced peaks in energy demand during the day. The Mōtū River, in the Gisborne region, was identified as having potential to generate a considerable amount of power, along with the Mohaka, in Hawke's Bay, and Whanganui rivers.

At the same time as this rush of enthusiasm for hydro generation, there was emerging political interest in protecting some rivers. In 1977, then Environment Minister Venn Young asked the Commission for the Environment to investigate a waterways protection policy. This resulted in the production of a discussion paper, *Wild and Scenic Rivers Protection*, which was released in December that year. In his foreword to the paper, Minister Young outlined the situation.

> In recent years the construction of hydro electricity and irrigation works have created an undercurrent of concern that development activities should be balanced by firm protective measures for certain New Zealand rivers, or parts of rivers, which offer outstanding scenery or recreational experience. This concern is sharpened by recent forecasts that New Zealand must expand its hydro electric capacity in a bid to reduce dependence on oil-fired power stations and to postpone the need for nuclear stations.

In the discussion paper, the Commission for the Environment highlighted the need to provide legislative backing for wild and scenic river designation, but it did not go so far as to recommend a specific mechanism to achieve this. The commission also suggested that responsibility for the designation process, along with the co-ordination of management and protection of designated rivers, should rest with a single agency. These tentative suggestions were put out for public feedback.

Some 110 submissions were received on the discussion paper, highlighting the degree of public interest in the matter. Several options for legislative change were suggested, including incorporating new provisions into existing legislation, or developing specific legislation

Previous pages: The Mōtū Falls. ROB SUISTED/NATURESPIC

Facing page: Rakaia River and Mathias River (centre) draining Kā Tiritiri o te Moana the Southern Alps and Rolleston Range. The Rakaia River is one of the largest braided rivers in the country and is fed by the meltwater of the Lyell and Ramsay glaciers. ROB SUISTED/NATURESPIC

Mōtū River mouth. Interest in the river for hydro generation stemmed back to the 1950s with eight potential dam sites identified. ROB SUISTED/NATURESPIC

for wild and scenic rivers similar to that in the United States.

After considering the various options, the commission recommended that existing planning and decision-making processes be used. The recently established Queen Elizabeth II National Trust (founded by statute in 1977 to protect open space for the benefit of the public) was identified as the best agency to have overall responsibility for a wild and scenic rivers policy. The commission also highlighted the need to protect both land and water, necessitating the alignment of catchment management with river protection.

It took two more years before the government officially committed to the protection of wild and scenic rivers. In a December 1979 joint statement, Environment Minister Venn Young and Works Minister Bill Young stated that 'rivers and sections of rivers that have outstanding wild, scenic and other natural characteristics should be protected'. By that time, however, the Ministry of Works and Development had already applied for water rights to investigate the Mōtū River for hydro development.

Former acclimatisation societies and Fish and Game national chief executive Bryce Johnson recalls that the wild and scenic rivers legislation was a quid pro quo for the National Development Act. 'We argued that the National Development Act enabled resources to be used so we needed one that enabled important things to be protected. Bill Birch acknowledged the need to balance the law and he became a supporter of the wild and scenic rivers legislation.'

It turned out that Bill Birch, who was Minister of Energy at the time, was something of a greenie, being particularly fond of nature.

In order to prod things along a Save the Rivers Campaign was launched in March 1981 at a conference co-sponsored by the New Zealand Canoeing Association, Federated Mountain Clubs and New Zealand Federation of Freshwater Anglers. The campaign sought to promote legislation that would enable rivers to be protected. As Gary Taylor recalled, 'we needed to get a positive initiative into the law to protect something, and wild and scenic rivers was the thing'.

At that time, the Ministry of Works and Development was opposed to special-purpose legislation that would enable rivers to be preserved in their natural state over the long term. The ministry was a champion of the 'multiple use' concept, which would enable rivers to be utilised for a range of activities, including economic development such as hydro alongside recreation. But there were others in government who argued that outstanding rivers, or stretches of them, deserved special protection in the same way as areas of land were conserved. This contest of ideas resulted in two competing draft Bills circulating around government corridors and vying for Cabinet approval.

The first Bill, produced by the Ministry of Works and Development, enshrined the multiple use approach, leaving it to catchment boards through the development of water allocation plans to provide protection for outstanding rivers. The ministry had also sought to incorporate a fast track for Crown water rights, which would be granted without the normal appeal rights to the Planning Tribunal. The second Bill, which was produced by a special interdepartmental committee, tasked the Minister of Works with initiating protection of 'nationally protected rivers'. The minister's proposal would then be subject to public submission before being finalised through an Order in Council.

As the Save the Rivers Campaign noted in its June 1981 *Rivers Report*,

> Both Bills have weaknesses and strengths. The 'multiple use' draft at least allows members of the public and organisations to initiate the rather weak protective provisions. The Order in Council approach leaves it solely in the hands of the nation's chief river dammer, the Minister of Works, to initiate the preservation of an outstanding river.

The report concluded, 'This is rather akin to giving the Minister of Mines the sole task of designating National Parks.'

The Ministry of Works and Development was given the task of stitching together a draft compromise Bill. This emerged on 1 September 1981 when the Water and Soil Conservation Amendment Bill was introduced into Parliament. It was passed in a rush, on 22 October, just prior to Parliament rising for the general election and only seven weeks after its first introduction. On behalf of EDS, Ian Cowper and Gary Taylor spent a week in Wellington lobbying, but were told that it was a matter of passing the Ministry's Bill or nothing. As a result of the exercise, they formed the view that there was an 'appalling lack of understanding of environmental issues on the part of most MPs'.

The new law was compromised from the start. There were tight restrictions around who could apply for protection. The final decision-making power was given not to the minister charged with looking after the environment, but to the minister in charge of damming rivers. Applications for water conservation orders were first to be heard by the National Water and Soil Conservation Authority, an entity chaired by the Minister of Works.

There was an ability for parties to appeal the authority's decision to the Planning Tribunal. However, although the tribunal was tasked with undertaking an inquiry into the application, it was not able to make a final decision. This also rested with the Minister of Works. Water conservation orders did not necessarily provide permanent protection, as they could be revoked simply by reversing the process that had led to their establishment. In addition, the scope of a water conservation order was constrained. It could only apply to the water in the river, with no ability to control the use of adjacent land.

EDS thought that the new statutory provisions were far from satisfactory, but the society was determined to make the new legislation work. As Gary Taylor observed at the time, 'It is excessively bureaucratic, unnecessarily complex, and above all provides a form of protection that can be undone at any time ... The flaw is simply that the nation's biggest dam builder [the Minister of Works] makes the final decision.'

The first river EDS and others wanted to see protected was the Mōtū. This was to provide a test case for the new legislation.

THE MŌTŪ RIVER flows between the Huiarau and Raukūmara ranges in the Bay of Plenty, reaching the sea some 41 kilometres north of Ōpōtiki. The middle reaches of the river flow through old-growth forest within the Raukūmara Conservation Park. The Mōtū has been described as one of the country's most beautiful rivers, with its rugged isolation, numerous rapids and cascades, and virgin bush flanking its banks.

The New Zealand Canoeing Association had been campaigning for the preservation of the river since 1975. By the early 1980s, the Mōtū River was regarded by many conservation and recreational groups as the last truly wild river remaining in the North Island. But the features that made it so attractive to canoeists, the steep narrow gorges with their high walls and numerous rapids, also made it suitable for hydro development. For this reason, conservationists had made little progress in protecting the river.

Interest in the Mōtū for hydro development stemmed back to the late 1950s. By the early '60s, it was concluded that developing the river for hydro would be more expensive than alternative thermal generation. But when the cost of oil increased during the 1970s, hydro generation on the Mōtū came back into the frame. It would likely involve the construction of four 100-metre-high dams cascading down the river. Eight potential dam sites had already been identified.

In 1978, the Minister of Works applied for

The Mōtū River flowing out of the Raukūmara Range. The Planning Tribunal recommended that virtually the entire river be protected, including the portion shown here. It was a major win for EDS and the conservation movement. ROB SUISTED/NATURESPIC

water rights to investigate hydro generation on the Mōtū. This was somewhat after the fact, as investigations on-site had already started a year earlier, prior to any authorisation. The water rights were duly granted by the National Water and Soil Conservation Authority in December 1979, without any public notice.

The decision to grant the rights, and the opportunity to appeal the decision to the Planning Tribunal, was only made public in one small advertisement in the *Gisborne Herald*. The statutory 28-day appeal period ran over the Christmas and New Year holidays, when most New Zealanders were away on leave.

Fortunately, someone from the New Zealand Canoeing Association saw the notice and the association filed an appeal with the Planning Tribunal. At the subsequent hearing, the association complained about the inadequate public notice. The tribunal ultimately concluded that, not only had the notice been inadequate, but the map references for the application were also incorrect. They referred not to the river at all but to a small area in the Raukūmara Range some kilometres away. As a result, the water rights were declared void.

A year passed before a new application for water rights was filed. This time, it was publicly advertised from the outset, and objections were heard by the National Water and Soil Conservation Authority in September 1981. At the hearing, EDS argued that consideration of the water rights should be deferred until the wild and scenic rivers protection amendment had been passed into law, and the status of the river under those provisions decided. But the authority dismissed the argument and granted the water rights in February 1982. EDS lodged an appeal to the Planning Tribunal, as did 93 other parties, indicating the level of now-mobilised opposition to the prospect of damming the river.

By this time, the new wild and scenic rivers statutory provisions had come into force, and EDS and others were pushing hard to get the river protected before it could be developed. There was a new Environment Minister, with Dr Ian Shearer taking over from Venn Young after the 1981 General Election. Shortly after Dr Shearer's appointment, Gary Taylor persuaded him to undertake a short rafting trip down the Mōtū River so he could experience its wildness and beauty for himself. The Commissioner for the Environment Ken Piddington and Rotorua East MP Paul East also came along for the ride.

The river visit generated significant conservation dividends. After the rafting trip, Shearer said he would oppose any development of the Mōtū; 'there is no room for compromise, one dam is one too many.' Piddington wrote to the Ministry of Works and Development stating that he would strenuously oppose any further investigations of the river.

By this stage, the enthusiasm of the Ministry of Works and Development for the hydro project was waning. After further investigation, the economics of harnessing the Mōtū for power generation appeared marginal. The narrow river gorges meant that the lake storage area created by the prospective dams would not be large. New Minister of Works Derek Quigley eventually agreed to adjourn the Planning Tribunal appeals on the water rights until the water conservation order application could be resolved.

There was now some urgency to get the Mōtū water conservation application lodged while the politics around the river's future was supportive. Under the new legislative provisions, EDS did not qualify as a party which could lodge an application. So Taylor got hold of Sir Thaddeus McCarthy, who had been appointed Chair of the Queen Elizabeth II National Trust. Sir Thaddeus was a jurist who had a great love for the New Zealand outdoors. Taylor asked him to lodge the Mōtū application in the trust's name. McCarthy agreed, but only on the basis that EDS would provide pro bono legal representation for the trust at the application hearings.

To support the application, EDS decided to review all of the information that was available on the river. Dr Stella Penny, who was at that time wife of EDS director Evan Penny and later went on to be the first woman put in charge of a regional conservancy in the Department of Conservation, wrote the report. Titled *Mōtū: A Wild and Scenic River*, it brought together a wide range of information about the features and values of the river. It concluded that the river should be preserved as 'wild and scenic'.

Over the 1982 Easter weekend, a group of EDS directors and supporters decided to take a commercial rafting trip down the Mōtū River to gain a first-hand experience of the river's beauty and to help with the society's efforts to protect it. But the trip was to take a tragic turn.

The Mōtū River flows through the rugged Raukūmara Conservation Park and has been described as one of New Zealand's most beautiful rivers. LLOYD HOMER, GNS SCIENCE

The EDS party camps on the bank of the Mōtū River as the weather deteriorates. Unbeknown to the group, the remnants of Tropical Cyclone Bernie were about to dump heavy rain in the high country above one of the river's main tributaries. EDS

It was an exuberant group of EDS directors, lawyers, family and friends that assembled on Thursday 8 April 1982 to embark on a four-day rafting trip down the Mōtū River. There were 17 in the group: Graham and Hilary Pettersen, Graham and Leith Wallace, Ian and Karilyn Cowper, Gary Taylor, Colleen Pilcher, John Timmins, Ian Lowish, Paul Cavanagh, Warren Templeton, Murray Cooper, Kevin House, Max Slimmer, Geoff Wood and William (Bill) Holt.

The trip commenced with a bus drive into the upper catchment of the river. Prior to the group's departure, the weather forecast had predicted cold conditions but was otherwise favourable. No outside communications were possible once the party reached the remote and rugged mountainous area where they were to join the river. Unbeknown to them, the remnants of Tropical Cyclone Bernie were about to dump heavy rain in the high country above the Mangaotane Stream, a tributary which meets the Mōtū just above the Ministry of Works and Development campsite at Opiti. The Ministry had identified the 'magnificent gorge' at Mangaotane as a potential location for a dam, so it had driven a road through to Opiti and brought in 12 huts for use by workers investigating the hydro-electric potential of the river.

Unaware of the danger ahead, the EDS group headed onto the river at around 3.30pm. The party was split between two rafts, each captained by a commercial rafting operator. It was slow going as the river level was low and after about an hour and a half they called it a day and camped on the riverbank.

By the next morning rain had set in, then the wind got up and the temperature plummeted. It was now bitterly cold. Notwithstanding the deterioration in the weather, the party pressed on, departing in the rafts at around 8.30am. It was still slow going due to the low river flow, with passengers exiting the rafts several times and carrying them over the shingle banks. One of the rafts developed a slow leak, so the party stopped at a campsite on the riverbank for lunch while attempts were made to repair it. Everyone was very cold by this point, but Bill Holt was feeling it more than most. He was the only member of the party without full-length wetsuit leggings.

The party set out again in the early afternoon. They had planned to camp for the night about an hour's travel downstream, but when they arrived at the site after only 30 minutes rafting, the two captains conferred and decided to press on to the Opiti campsite. This was to be a fateful decision. To get there the rafters had to pass through the very steep-sided rocky gorge where the Mangaotane Stream met the Mōtū. Once they entered the gorge, there was no way back out. They had to carry on through before reaching the Opiti campsite at the other end.

Ian Cowper remembers entering the gorge, drifting around the corner of the river, and then seeing a huge wall of brown water in front of the rafts. 'When it hit us, it was so extreme that we bounced in and out of the raft several times until we all got washed out. I got separated from my wife and spent 24 hours wondering what had become of her.'

Gary Taylor also recalls entering the gorge and seeing a huge wall of water spilling out of the Mangaotane Stream. It was running right across the path of the rafts before hitting the high rock wall on the other side of the main river and splashing back. 'The raft flipped over and I was in the water. I wasn't doing great. I bumped into the other raft that was upright. Ian Cowper, amongst others, was in it. He looked at me and I looked at him. I said "help" as he was a bit frozen in shock. We were all in shock. He then grabbed me and pulled me into the raft.'

A 5-metre-high wall of water had hit the rafts side on. Both rafts were pounded, and they flung their occupants into the turbulent and freezing cold water. Several people managed to cling onto the steep rocky sides of the gorge. They then clambered up through lose shale to get clear of the raging waters. Others managed to get back onto the rafts, which then floated out of the gorge. While unsuccessfully trying to rescue a party member who was left clinging to a rock face, one of the rafts was punctured.

The shocked and battered rafters warm up around a fire at the Opiti campsite with the ex-Outward Bound party that was fortunately camping in the area and managed to raise the alarm. EDS

The first raft successfully made landfall adjacent to the Opiti huts, where an ex-Outward Bound party was camping. They helped warm up the rafters and shared their limited food. One of the campers ran four hours through rugged country to reach the Ōpotiki Police Station and raise the alarm.

The next day the revived rafters set up a 'river watch' in an attempt to rescue anyone who might be washed down the river from the gorge. Kevin House floated past after re-entering the river that morning. He had sheltered in the gorge overnight where he had been hit by a rockfall which injured his knee. The river watchers threw a rope and managed to bring him safely ashore.

The second raft, which had been ruptured, was held in the grip of the current and carried on downstream past Opiti. It was only around dusk that the five occupants managed to make landfall. Their captain had earlier been separated from the raft while

Above: Kevin House (far left), Colleen Pilcher (third from left), Paul Cavanagh (fifth from left, standing) and Ian Lowish (sixth from left). EDS

Above right: Ian Cowper (in the blue wetsuit to the right of the image) chatting to the ex-Outward Bound campers. EDS

unsuccessfully trying to secure it to the shore. Freezing cold and exhausted, the five survivors dragged the raft up out of the water, turned it over, and made camp under it for the night. It was at this point that they saw Bill Holt floating down the river. They were unable to reach him and he was washed out of sight.

Lawyer Geoff Wood was one of the rafters who was left behind in the gorge. After being tipped out of the raft he had been carried downstream in huge waves amid jostling logs. Just as darkness was falling the water current swung him to the shore at the bottom of a steep shale bank. He spent over an hour clambering up the bank in the driving rain before he found shelter under a tree. He got little rest however. In the darkness he could hear the rocks sliding all around him and at one point a large boulder hit the tree he was sheltering under.

In the morning, Wood was cold and exhausted, and he found it very hard to decide what to do. He later described to a *New Zealand Herald* reporter, 'I knew no one would come down the river. I thought that if I was going to die I didn't want to die slowly from exposure. I reasoned that I had a chance if I got back into the river.' So, courageously, Wood re-entered the river. As he recounted, 'the waves were huge. I was out of control. You couldn't tell if you were up or down. I was spinning around all the time.'

Wood eventually floated out of the gorge and past the Opiti campsite. He was seen by Gary Taylor, Ian Cowper and John Timmins, who were on river watch. They threw him a rope but missed by about 3 metres. As Geoff recalled, 'I struggled to side

Right: Using a radio to co-ordinate the rescue of the rafters stranded in the gorge and on the other side of the river. Gary Taylor is on the far left, John Timmins is third from the right and Karilyn Cowper is on the far right. EDS

stroke to the shore at Opiti. They threw a rubber ring but I was going too fast. I did not catch it. My heart sank seeing the rescue so close but not being able to do anything to get there. I thought I was going to die.

'Then there was a shallow piece of the rapids. I dived for the shore. I sat on the shore for a while. I thought I would be a lot better off if I kept moving. I walked along the bank for another 100 yards. Then I saw two red lifejackets hanging up in a tree. That was the most wonderful moment of my life.' By chance Geoff had stumbled on the camp set up by the occupants of the second raft. He was later helicoptered out to hospital to be treated for hypothermia.

Ian Cowper's wife Karilyn Cowper, who had been thrown out of the raft in the gorge, managed to clamber to shore along with Ian Lowish. Lowish was in a bad way with a face injury so Cowper helped him into the bush where they both sheltered. Colleen Pilcher, who had been spilled out of the raft just upriver of the gorge, climbed up the rock face and made her way downriver through the bush. She fortuitously stumbled across Cowper and Lowish the next morning. The group heard a helicopter fly up the gorge, managed to attract its attention, and were picked up and taken to the Opiti camp site.

Eventually all the rafters were safely rescued except one. The body of Bill Holt was found four days later, 19 kilometres north of Whakatāne, and more than 50 kilometres from the mouth of the Mōtū River. Holt was a book dealer from Tāmaki Makaurau Auckland who held a doctorate in economic history from the University of Chicago. He was just 35 years old.

Notwithstanding the shattering experience of the EDS rafters described in the panel text, the application to protect the Mōtū was lodged with the Ministry of Works and Development a couple of weeks after the trip, on 30 April 1982. It sought to protect the river along its entire length, from the Mōtū Falls high up in the Raukūmara Range all the way to the sea.

Then nothing was heard. The application disappeared into the bowels of the bureaucracy. In June, Taylor wrote in the EDS annual report that the ministry might be dragging the chain as he believed many of the department's officers do not 'support the new legislation and could be adopting an unco-operative attitude towards processing the Queen Elizabeth II Trust's application'.

It was not until December 1982 that the application was heard by the National Water and Soil Conservation Authority. The authority had received 222 submissions, including those from the Ministry of Works and Development, 56 statutory bodies including local councils and 156 individuals. In addition, 11 petitions signed by 254 people had been forwarded. An overwhelming 214 submissions favoured making the order to protect the river, indicative of the depth of support for the proposal.

Opposition came mainly from the local councils that favoured the development opportunities provided by hydro power. Opotiki County Council objected to making the order and the East Cape Catchment Board, East Cape United Council and Poverty Bay Electric Power Board all wanted the matter deferred in light of the benefits of hydroelectricity generation to the region and the need to complete the regional planning scheme.

The National Water and Soil Conservation Authority issued its decision in February 1983. It concluded that only the middle portion of the river deserved protection. This was because the upper and lower sections lacked the same outstanding characteristics. Even then, the proposed protection was only partial. The authority determined that provision should be made for the granting of water rights associated with hydro-electric investigation, soil conservation works and wild animal control.

On hearing of the decision, Gary Taylor commented to *The Dominion*, 'The authority wouldn't recognise conservation if they tripped over it, they're just developers. The decision highlights the need to get the water authority out of the Ministry of Works. We can't have dam builders making conservation decisions.'

EDS had two main issues with the authority's decision. First, it only protected part of the river, and therefore would have enabled a hydro scheme to be developed in its upper reaches, and, second, it enabled water rights to be granted for hydro investigation. In EDS's view, the standard applied by the authority was 'still wedded to the multiple use ethic and applied such a strenuous test in assessing the wild and scenic values of the river that other rivers do not stand a chance of being protected'. It was therefore crucial that the authority's approach be overturned on appeal.

The Planning Tribunal heard the appeals against the authority's decision in October 1983. The five-day hearing, chaired by Judge John Treadwell, was held in Whakatāne. Ian Cowper and Gary Taylor appeared for EDS. As agreed, EDS also provided pro bono lawyers for the Queen Elizabeth II National Trust. They were EDS founder David Williams and his colleagues Mike Holm and Simon Berry from Russell McVeagh.

At the hearing, the Ministry of Energy confirmed that the Mōtū had been dropped from the Energy Plan, and it was not likely to be reconsidered until the much more economic options of coal burning and South Island hydro projects had been pursued. The high cost of developing the river had worked in favour of preserving it. The Ministry of Works and Development also did not oppose the application, accepting that the river warranted protection, although seeking the right to continue certain hydro investigation work. The regional council and several other local authorities still argued that the application should be deferred.

David Williams and his team contended, on behalf of the applicant, that the National Water and Soil Conservation Authority had applied the wrong legal test in concluding that all parts of a river covered by a water conservation order needed to qualify as outstanding. Instead, they argued, the river should be 'taken as a whole' when considering whether it should be preserved as far as possible in its natural state. The Planning Tribunal ultimately agreed that this interpretation was within the spirit and intent of the Act. It set an important precedent for other rivers.

In concluding its submission, EDS told the Tribunal:

> This is the first time EDS has participated in statutory procedures where a conservation initiative has been taken. We respectfully ask you to approach your recommendations with a generous disposition towards the conservation ethic. The legislation provides for rivers

The lower reaches of the Rakaia River flowing through the Canterbury Plains to the sea south of Christchurch. LLOYD HOMER, GNS SCIENCE

> or parts of rivers to be protected and we ask you to protect all of the Mōtū — from the Falls to the Sea. To do so would be consistent with the scheme of the Act, would give recognition to deep-seated aspirations that are widespread in the community, and in the final analysis would preserve all development options for the future.

When the tribunal issued its recommendations in January 1984 it was evident that it sympathised with the views expressed by supporters of the water conservation order. The tribunal recommended that the river be protected from the Mōtū Falls right down to the bridge on the coastal road, only excluding the small stretch of water from the bridge to the coast. It also recommended expressly prohibiting the granting of any water rights for dams or hydro investigations.

EDS was delighted with the outcome: 'the fact that the Planning Tribunal recommended protection for the Mōtū River has cheered conservationists throughout the country. It is the first time that we have been able to take a conservation initiative through the Courts and we have won.'

The Minister of Works still had to make the final decision on the water conservation order, but given the ministry's positive stance at the hearing and strong public support for protection, this happened as a matter of course. The water conservation order was gazetted on 7 February 1984, two years after the application was first lodged.

David Williams later recalled,

> I rank as one of the greatest achievements of EDS the 1984 protection order on the Mōtū River, a case in which I was lead counsel for EDS who took the case for the Queen Elizabeth II National Trust. I learnt a lot about wildlife in that case and especially the peculiar characteristics of the rare blue duck. One of the scientists said to me it has a 'crepuscular disposition' — I had to run to the dictionary to find what it meant — that its major activities took place in twilight hours, a word I have since found useful when applied to some of my friends.

—ooo—

THE NEXT WATERWAYS off the block for protection were the Rakaia and Ahuriri rivers. The Rakaia is one of the largest braided rivers in the country. It first emerges high up in the Southern Alps, being fed by meltwater from the Lyell and Ramsay glaciers. The river runs some 150 kilometres through the foothills of the alps and over the Canterbury Plains before reaching the sea approximately 50 kilometres south of Christchurch. The Ahuriri River is shorter, flowing some 70 kilometres along the southern edge of the Mackenzie Basin, before entering Lake Benmore near Ōmarama.

This time others took the lead in seeking river protection and EDS provided support. The acclimatisation societies applied for the Rakaia River water conservation order and the Wildlife Service applied to protect the Ahuriri River. The initial hearings for both applications were held in late 1983.

The Rakaia River application was hotly contested as farmers on the Canterbury Plains

The Ahuriri River looking upstream towards the Ahuriri Valley. The braided Ahuriri flows around the edge of the Mackenzie Basin and provides habitat for the critically endangered black stilt. ROB SUISTED/NATURESPIC

The Ahuriri River looking upstream towards the Dunstan Range. The water conservation order to protect the natural state and minimum flows of the river was confirmed in 1990. ROB SUISTED/NATURESPIC

were keen to extract river water for irrigation. The river also provided some of the best recreational salmon fishing in the country, which was why the acclimatisation societies were keen to protect it. The application ended up in the Court of Appeal after Federated Farmers successfully challenged the Planning Tribunal's recommendation to grant the order in the High Court. David Williams and Simon Berry acted for EDS.

The case focused on the correct application of the wild and scenic legislative provisions. Federated Farmers argued that these required a balancing exercise and there was no inbuilt preference for conservation interests. Each case should be weighed up on its particular facts without any general presumption, it contended. In contrast, EDS and the acclimatisation societies argued that there should be a presumption in favour of conservation, based on the special objective for water conservation orders inserted into the legislation, which was 'to recognise and sustain the amenity afforded by waters in their natural state'.

The latter approach was adopted by the Court of Appeal with the judgement of Justice Robin Cooke emphasising that a compelling case was needed to overturn the presumption in favour of conservation. The judgement stated,

> as a general working rule or guideline preservation of the natural state, either as fully as possible or to the extent of protection of outstanding characteristics or features, is to be aimed at unless clear and sufficient reason is shown to the contrary. The ultimate criteria must be the public interest. The presumption is in favour of conservation. A strong, really compelling case is needed to displace it.

The case established a significant legal principle which helped ease the way for the creation of future water conservation orders.

The Rakaia River water conservation order was hard-fought and it was finally confirmed in October 1988. The order protected the upper parts of the Rakaia River and its tributaries in their natural state and provided for minimum flows between the Rakaia Gorge bridge and the sea. Existing water takes were allowed to continue and provision was made for the Lake Coleridge hydro power station to carry on its operations. In 2023, EDS was back in court alongside Fish and Game, seeking to protect the integrity of the order.

The Ahuriri River water conservation order was less contentious. One of the arguments in support of the order was that the river was the habitat of the rare and critically endangered black stilt, of which there were thought to be only 80 left in the world at the time. The birds nested in the braids of the river. Simon Berry, who acted for EDS on the application, recalls that at the start of the hearing in Ōmarama there were some references to shooting all the black stilts so the water conservation order wouldn't be needed.

Gary Taylor also attended the hearing on behalf of EDS. He recalls,

> I think by the end of the hearing there was more of a meeting of minds on the need for a water conservation order from members of the farming community who initially opposed it. One of the farmers offered to give me a lift when the hearing concluded. It turned out the lift was in his little Cessna with hay bale remnants strewn across the floor and bale twine everywhere.

The Ahuriri water conservation order was finally confirmed in July 1990. Similar to the Rakaia order, it protected the bulk of the river and its tributaries in their natural state and provided for minimum flows in the Ahuriri River itself. The water conservation order legislative provisions were subsequently carried over into the Resource Management Act 1991 and there are now 15 rivers and lakes protected by them.

Former Fish and Game national chief executive Bryce Johnson has been in the thick of seeking protection for rivers. He later observed, 'EDS always stayed in the litigation space. It was a new thing for Fish and Game and a risky thing. Having EDS there as a fellow activist, at that level, made a huge difference. We got the legislation and EDS got the first application for the Mōtū. Fish and Game then followed through with eight or ten others and is still going. It was pretty good stuff.'

5. Into hot water

IN MARCH 1979, the Papakura Geyser at the Whakarewarewa geothermal field in Rotorua failed to blow. The geyser was well-known for its spectacular and continuous hot-water eruptions, which had typically reached heights of 2 to 3 metres. But now it was silent. It had followed the fate of two nearby geysers (including the Waikite Geyser which once sent hot water up to 20 metres into the air) and served to highlight the ongoing decline in geothermal activity in the area.

AUCKLAND UNIVERSITY physicist Dr Ron Keam was an expert in geothermal processes and he, along with Rotorua-based geologist Ted Lloyd, expressed alarm about the decline in the Rotorua geothermal field. This caught the attention of EDS director Dick Bellamy. When Bellamy heard about the issue he was keen for EDS to get involved.

Bellamy remembered being taken as a child to the Wairakei Thermal Valley when it had been an active thermal area full of geysers and 'a marvellous place to go'. But since the establishment of the Wairakei geothermal power station in 1958, it had become a silent moonscape. It provided a vivid illustration of what could happen to surface thermal activity when large amounts of geothermal water and heat were extracted and not returned to the aquifer.

The Whakarewarewa valley, at the southern end of the Rotorua geothermal system, once contained the most concentrated collection of active geysers in the world, and had been a major tourist attraction for well over a century. But after the government wrested control of the thermal area from Māori during the 1880s, facilitated by the Thermal-Springs Districts Act 1881, the city of Rotorua was built bang on top of the field.

Rotorua was originally a small spa town, established during the 1880s to cater for tourists visiting the geothermal wonders and enjoying the thermally heated bathing pools. But it had grown substantially since then and by 1979 the city had close to 50,000 residents.

One of the attractions of living and doing business in Rotorua was the availability of cheap geothermal energy for heating and bathing. Driving bores into the ground to extract geothermal heat was common practice, being lightly regulated and largely unmonitored. Special legislation, in the form of the Rotorua City Geothermal Energy Empowering Act 1967, had provided a carve-out for the city whereby the Rotorua City Council (which became Rotorua District Council in 1979) managed the exploitation of geothermal energy in place of the Minister of Energy. This had enabled the council to liberally grant bore licences with little scrutiny.

For a period, Rotorua residents were even encouraged to sink bores to tap geothermal water and heat. As Roger Brewster, a Rotorua lawyer and chairman of the Geothermal Users Association explained, during the 1960s 'electricity was short and oil became an expensive commodity. We were doing the country a good turn at the time.'

By 1979, when the Papakura Geyser failed, there were more than 400 geothermal bores operating in Rotorua. About 250 of these belonged to private residents, 60 to government and other institutions and a further 90 were operated by motels, hotels and other businesses.

There was no charge for the use of geothermal energy in Rotorua and much of the extracted water was not reinjected into the aquifer and heat was wasted; one estimate put the loss at some 80 to 90 per cent. Bore valves were typically left wide open year-round, belching out steam into the open air. Many pipes lacked any insulation and the temperatures in offices and hotels was often uncomfortably hot. No one knew how much geothermal fluid was being taken from the field overall as no monitoring was undertaken and there were no reporting requirements on users.

Previous pages: Roto-a-Tamaheke lakelet at Whakarewarewa. RAEWYN PEART

Whakarewarewa, Rotorua, in the early 1900s showing only a small cluster of tourist development near the geothermal area. AUCKLAND LIBRARIES

As the Rotorua urban area had expanded, there had developed a complex mosaic of provisions under which the geothermal resource was supposedly managed. In the old parts of the city, which came under the empowering legislation, geothermal use was under the ambit of the Rotorua District Council. But in the newer parts of the city, geothermal use was managed by the Minister of Energy under the Geothermal Energy Act 1953.

The biggest user of geothermal energy by far was the government-owned Forest Research Institute. It had been drawing geothermal energy since 1953 to heat its glasshouses as well as offices and a training centre. By 1978, the institute had four wells, drawing some 300–400 tonnes of geothermal fluid each day. These were some of the closest bores to the Whakarewarewa geysers, being located only around 150 metres away.

One of the wells had reached the end of its useful life, so in 1979 the Forest Service asked the Ministry of Works and Development to drill a new one. This was proposed to be 500 metres deep, double the depth of the existing wells, in the hope of finding higher temperatures. The institute campus was outside the area covered by the council empowering legislation, so a licence was required from the Ministry of Energy.

Geyser Flat, Whakarewarewa. The health of the Prince of Wales and Pōhutu geysers, along with other geothermal features at Whakarewarewa, was left in the hands of the Rotorua District Council, which liberally granted bore licences with little oversight or monitoring. RAEWYN PEART

This was duly granted in December 1979, outside the public eye, as the Geothermal Energy Act did not make any provision for public participation in decision-making.

The new well also needed a water right. The Forest Research Institute made an application for this in December 1979, along with its existing wells which, having been established prior to the passage of the Water and Soil Conservation Act 1967, now also required water rights under the new legislation.

There had been some legal uncertainty as to whether the new water and soil legislation actually applied to the taking of geothermally heated water. The Bay of Plenty Catchment Commission, which was in charge of processing applications for water rights in Rotorua and the wider region, had received legal advice that the Act did not apply. It was refusing to process water rights for geothermal energy. But the Ministry of Works and Development took a contrary view and considered water rights to be necessary.

Despite the catchment commission's doubts about the need for water rights, it did process the Forest Research Institute's application. This was duly advertised for public comment, but none was forthcoming, highlighting the lack of public engagement with such legal processes at the time.

Notwithstanding the apparent lack of

public interest in the proposal, the catchment commission was well aware of the loss of the Papakura Geyser and the broader concerns of scientists and Māori about the health of the Rotorua geothermal field. It therefore decided to hold a hearing, to which it invited experts from the Ministry of Works and Development and the Department of Scientific and Industrial Research (DSIR).

It became apparent during the hearing that rather than simply maintaining its current usage of geothermal energy, the institute had applied to double its take, a quantity 'considerably greater than needed'.

The commission took a dim view of this given the possible effect of a new deep well, coupled with much greater draw off, on the system, especially the fragile Whakarewarewa geyser field. It recommended that the water right not be granted and at this point the institute withdrew its application. This meant that not only did the institute not have water rights to permit the drilling of a new bore, its existing wells, which continued to operate, were also without the requisite rights.

It was at this point that EDS became engaged in the geothermal issue. As a start, and in order to draw public attention to the poor management of geothermal resources, the society decided to convene a seminar in Rotorua. This took place in October 1980, in association with the Nature Conservation Council. After canvassing a range of issues related to the management of geothermal resources, the seminar called for a moratorium on further development of certain geothermal fields until a national strategy for balancing conservation and the exploitation of interests was prepared. The seminar noted that 'political pressure from local residents who were determined to maintain their free energy supplies meant that conservationists' warnings were ignored'.

In order to keep up the political pressure, a second seminar was convened the following year. Titled 'Geothermal Systems: Energy, Tourism and Conservation', it aimed to increase public awareness of the values of geothermal systems, explore ways of obtaining a balanced use of them, and make recommendations to government concerning the establishment of appropriate policy. The seminar also called for the urgent establishment of an independent commission of inquiry to examine all aspects of the geothermal issue.

In a press statement following the seminar, Gary Taylor observed, 'it is abundantly clear that we must evolve a strategy by which we identify those geothermal systems we should preserve for their scenic and scientific values and those which can be exploited for direct use in industry and for electricity generation. Decisions have up to now been made on an ad hoc basis and in a policy vacuum.'

He went on, 'There are still outstanding questions that must be urgently resolved. Of special importance is the question of a moratorium on further development until the proposed national strategy has been formulated.'

Meanwhile, the government was taking steps to resolve the legal uncertainty over the application of the Water and Soil Conservation Act to geothermal activity. In an amendment to the Act, which was primarily intended to enable protection of wild and scenic rivers (as described in the previous chapter), a provision was slipped in to make it clear that water rights were required for taking geothermal water and heat.

The new provisions came into effect on 1 April 1982. In order to give operators of

The Roto-a-Tamaheke lakelet at Whakarewarewa. The water level of the lakelet dropped and hot springs around its edge began to behave unusually when the government-owned Forest Research Institute drilled a deep geothermal well just 150 metres away. RAEWYN PEART

geothermal bores adequate time to apply for the now clearly needed water rights, the catchment commission issued a general authorisation that legalised existing bores until 31 January 1983. This regularised the Forest Research Institute's existing bores for the time being. However, the institute also claimed that it authorised the drilling of the new replacement well. This was strongly refuted by EDS lawyers.

The institute had clearly been waiting for the general authorisation to be issued because once it came into effect on 1 April drilling quickly started on the replacement well. But the engineers soon struck problems. They had to drill deeper than expected to access the geothermal fluid, and while doing so they hit a permeable layer of sediment which allowed some of the fluid to disperse. In response, they pumped a large amount of cold water from the town supply into the bore hole to sluice it out.

Soon afterwards, in early July 1982, the water level at the Roto-a-Tamaheke lakelet dropped and some of the hot springs around its edges started to behave unusually. The lakelet was just 150 metres away from the institute's new bore. Ted Lloyd viewed the situation with the 'utmost concern' and saw it as 'a vivid illustration of the extreme fragility of the pressure regime' driving the springs. EDS wrote to the Minister of Forests Jonathan Elworthy

The Waikite Geyser at Whakarewarewa in 1892. This spectacular geyser was once one of the biggest attractions at Whakarewarewa but it stopped erupting in 1967. RYAN THOMAS, AUCKLAND LIBRARIES

asking him to close the bore down, but the minister declined to take any action.

The efforts of EDS and others to publicise the ongoing loss of geothermal features started having an effect. In June 1982, Energy Minister Bill Birch announced a moratorium on all drilling (except for monitoring purposes) in 11 geothermal fields managed by the government. But the moratorium did not apply to the area controlled by the Rotorua District Council.

An Official Geothermal Coordinating Committee was then formed by the government to prepare a discussion paper on the issue. In addition, a geothermal monitoring programme in Rotorua was finally to be established. The government had proposed such a programme some two years earlier, offering to pay the greater share of the cost, but the Rotorua District Council had until then refused to contribute anything. Although there was now some action, it was all too little too late. Nothing was being done to shut down existing bores. And there were more worrying signs that geothermal activity at Whakarewarewa was fading.

The Waikite Geyser had failed to come to life during conditions of above-average rainfall which would normally drive it into action. In February 1982, the Korotiotio Springs, which had been used as a source of water for the old oil baths at Whakarewarewa, ceased overflowing and the water-level dropped. And at the end of July 1982, the Parekohuru Spring, which had been known as the 'Champagne Pool' for its fizzy and boiling surges, and which had been one of the most stable features at Whakarewarewa, ceased discharging.

Although over-exploitation of the geothermal field was the most likely explanation for the loss of activity, many Rotorua residents and business people claimed that their bores had no effect at all on the health of the geysers and boiling springs. They variously blamed the lack of rainfall, a drop in the level of Lake Rotorua, or the realignment of the stream running through Whakarewarewa as the cause.

The thermal area had always been erratic, they argued, so there was nothing unusual in the current drop in activity.

But geologists working in the region saw things differently. Ted Lloyd wrote of the failures being 'unprecedented in the known history of the field'. Ron Keam highlighted evidence from overseas that indicated that geysers adversely affected by geothermal exploitation may not re-start, even when the exploitation ceases. The damage could be irreversible. In Lloyd's view the current situation could only be addressed by a significant reduction in geothermal draw-off. 'Bores must be shut off!' he expounded. Otherwise 'we will have to send our grandchildren to Yellowstone National Park to see a geyser'.

Responding to the concerns of the scientists, EDS increased pressure on the Rotorua District Council to shut bores. On 30 July 1982, lawyer Ian Cowper wrote to the council on behalf of EDS and Whakarewarewa landowner Andrew Rangiheuea expressing concern at the situation and urging the council to use its regulatory powers to close bores down. The letter argued 'the public interest requires your Council's intervention or one of New Zealand's priceless geothermal heritages will be lost forever.' Although the council acknowledged receipt of the EDS letter it failed to make a substantive response for some months.

The letter was accompanied by a series of press releases issued by Gary Taylor, which were taken up by the local press. In a press statement on 2 August, Taylor summed up the situation.

> We are witnessing the disappearance of one of New Zealand's prime tourist attractions. It is disappearing because of greed and shortsightedness – greed on the part of domestic users who cumulatively are responsible for the excessive draw-off and shortsightedness by the politicians, both local and national, who refuse to act.

Māori were especially concerned about the loss of their precious taonga and Sir Peter Tapsell, who was then MP for Eastern Māori, successfully promoted half a day's debate on the issue in Parliament in July 1982. During the debate he pointed out,

> Thermal areas are not rare – there are many throughout the world – but geysers are extremely rare. They occur in Iceland, Yellowstone in the United States, Tibet, Kamchatka, and New Zealand. The New Zealand field is amongst the best, and is the most easily accessible in the world. There was a time when New Zealand had 130 geysers, but now it has only 5, all in the Whakarewarewa field.

Tapsell argued for the establishment of a single geothermal energy authority which should be given the power to administer, monitor and manage the Rotorua geothermal system 'to ensure the steam is used wisely, and in such a manner not to damage the geysers'.

Meanwhile, given the national significance of the Whakarewarewa geysers, Energy Minister Bill Birch had decided to set up an urgent review into the management of the Rotorua geothermal field. The review was undertaken by officials from the Ministry of Works and Development, the Ministry of Energy and the DSIR. They released a report in late August which recommended the installation of more efficient heat exchangers, the closure of wells

Mud pools at Whakarewarewa. Despite the reduction in geothermal activity at Whakarewarewa, most bore owners indicated that they were not prepared to cut back usage and the Rotorua District Council declined to take any action. RAEWYN PEART

from December to January when they were not required, and a ban on any new wells within 1.5 kilometres of Whakarewarewa. The review also suggested that control of the system should be taken off the Rotorua District Council on the basis that 'at this time it does not have the necessary technical expertise'.

The unwillingness of the Rotorua District Council to address the situation was confirmed in its November letter of reply to EDS and Andrew Rangiheuea. The council made it clear that it was not planning to take any action on the issue, claiming that there was 'no specific scientific evidence' available to support EDS's claims and that it could be obliged to pay compensation to bore owners, the cost of which would be 'massive and crippling'.

As well as failing to take action to close existing bores, even just over the summer months when they were not required, as recommended by the government review, the council was continuing to issue new well licences (albeit outside the 1.5-kilometre radius of the Pōhutu Geyser), further exacerbating the problem. In the face of the crisis, the council appeared unable to change course.

As the surface activity across the system, including at Whakarewarewa, further declined, and the council took no perceptible action to implement the recommendations of the urgent review, Bill Birch set up a second group to investigate further. This time it was a joint task force between central government agencies and the council. The task force undertook a survey of bore owners and one of the questions asked was whether they would be prepared to make voluntary reductions in geothermal take. Despite the evident reduction in geothermal activity, and potential negative impacts on tourist numbers and consequently the city's economy, 70 per cent indicated that they would not be prepared to cut back usage.

RAEWYN PEART

Ian Cowper completed a Bachelor of Laws (Hons) at the University of Auckland and was admitted to the bar in 1971. His honours dissertation focused on riparian rights under old English law, which piqued his interest in water regulation. Cowper started his legal career at Morpeth Gould in 1971 and he became a partner there in 1975. He left the firm in 1991 to join Chapman Tripp for four years before moving to Bell Gully.

In 2002, Cowper established boutique resource management legal firm Cowper Campbell with Janette Campbell. A decade later the firm merged with Meredith Connell, where Cowper spent the last three years of his legal career. His specialty was planning and resource management law and he was the recipient of the Resource Management Law Association's Outstanding Person Award in 2014.

Cowper was an inaugural student director of EDS in 1971 and he sat on the board for 11 years, until 1982. He has supported many EDS cases during that time. Cowper became involved in EDS because of his love for New Zealand and its environment. As he observed, 'I am interested in the environment and protecting it and someone needed to stand up. It seemed it was always the last consideration rather than the first.'

When the task force issued its report in July 1983 it was pretty much business as usual. The report failed to tackle the issue head-on, focusing almost exclusively on efficiency gains rather than bore closure. As EDS stated in a subsequent press release, 'The Task Force Report is a major disappointment. It fails to address itself properly to the hard decisions that need to be made if Whakarewarewa is to survive. Bores must be closed down immediately. By not recommending that action, the Report has not significantly advanced the issue at all.'

It was 1986 before the government, through a Cabinet directive, finally took back control of the Rotorua geothermal field from the Rotorua District Council and initiated a bore closure programme. A further ministerial directive was issued in 1989. As a result, most bores within a 1.5-kilometre radius of the Whakarewarewa geyser area were closed and a licensing and charging regime put in place for those bores that remained open. There was to be no fluid withdrawal from this area and only down-hole heat exchangers (which extracted heat, but not fluid, from within the bore) were permitted. Residents and hoteliers put up stiff opposition to the plan, including a legal challenge in the High Court, but this was swiftly defeated. Over 100 wells were cemented up and around 300 bores closed.

Since that time, the management of geothermal systems has become a function of regional councils under the Resource Management Act 1991. The Rotorua system is now managed by the Bay of Plenty Regional Council under the Rotorua Geothermal Regional Plan, which has largely carried over and built on the ministerial directives of the

1980s. The 1.5-kilometre exclusion zone is still in place, only down-hole heat exchangers are allowed within the zone, and there is to be no increase in the taking of heat. Reinjection of geothermal water is a requirement across the entire system, except in limited circumstances.

The new management regime has resulted in substantial recovery of parts of the Rotorua geothermal system. Monitoring bore data has showed a rapid recovery of aquifer levels once the bores were closed. The height of the geothermal aquifer is now some 2 metres above pre-bore closure levels.

However, recovery of surface features has varied. Kuirau Park, a geothermal area close to the CBD, has leapt back to life, but features at Whakarewarewa have shown a more mixed response. The Parekohoru Spring resumed boiling overflow surges in 1989 and the Papakura Geyser erupted for the first time in 35 years in 2015. However, the Korotiotio Springs are in a state of flux and many springs associated with the Roto-a-Tamaheke lakelet are still not flowing. The Waikite Geyser has not erupted since 1967.

AT THE SAME TIME that Whakarewarewa was suffering from over-exploitation, the government was planning to develop a virtually undisturbed field near by. The Waimangu hydrothermal system lies 22 kilometres south-east of Rotorua at the south-western extremity of Lake Rotomahana. The field was formed during the 1886 Tarawera eruption, making it the only large hydrothermal system in the world where surface activity has commenced in historic times, and was directly due to a volcanic eruption. It features two large hot lakes: Frying Pan Lake (claimed to be the largest hot-water spring in the world) and Inferno Crater Lake, as well as numerous boiling springs and occasional hydrothermal eruptions. For four years, between 1900 and 1904, it contained the world's largest known geyser. It was named Waimangu (black waters) by local Māori, referring to the dark-coloured water full of rocks and mud regularly expelled by the geyser. The name was also used to refer to the wider geothermal area, which became a significant tourism attraction.

Ron Keam had spent years studying the field and was well aware of both its uniqueness and volatility. He was therefore very concerned when, in early November 1978, the Minister of Works applied for water rights to test the area to determine if it would be suitable for geothermal energy supply. If it proved to be so, the minister was planning to make geothermal energy available to a timber-processing plant being established on the northern slopes of Maunga Kākaramea (then known as Rainbow Mountain), just south of Rotorua. The plant was operated by a company called F J Ramsey Limited.

The National Water and Soil Conservation Authority duly granted the water right application. As EDS director Dick Bellamy recalls 'Ron had seen that EDS were folks that did things so he approached us about the application.' EDS helped Keam appeal to the Planning Tribunal and EDS director Ian Cowper ran the case. EDS was unable to lodge proceedings in its own name as it did not meet the standing requirements for appealing against Crown water rights.

As Ian Cowper recalls,

> government wanted to drill some test bores near Waimangu and it was going to interfere with Ron's historical observation

> of a particular system there. There was a lake with a long throat. The water level would rise and then disappear down the throat. Ron would predict when it would come back up, form a lake and bubble for a while, before disappearing again. He had been studying that for ages.

Development of the system could disrupt his future research, as he would not know if the observed behaviour was a result of natural forces or the physical interference with them.

The Planning Tribunal hearing was held in September 1979. Cowper forcibly argued, in front of Chief Planning Judge Arnold Turner, that the scenic and scientific values of the Waimangu field were such that they should not be placed at risk by permitting draw-off or exploration and ultimately use by a timber-drying plant. Waimangu was the only major geothermal field left in New Zealand which had not been affected or modified in some way by human intervention.

It was a persuasive argument and the tribunal upheld the appeal. In a subsequent workshop organised by EDS and the Nature Conservation Council, Judge Turner provided more background on the decision.

> Our conclusion from the evidence and submissions was that the benefit which may follow from the exercise of the right sought was not sufficient to justify the detriment which might be caused to the scenic and natural features of the Waimangu thermal area ...
>
> We did not say that the scenic and natural features of the Waimangu area must be forever protected and that no right should be granted to take hydrothermal fluid from the area. We said that if it is desired to explore the energy potential of the Waimangu and Waiotapu areas, then we would expect that a comprehensive plan of exploration would first be prepared and that a decision to explore would not be made without full evaluation of the likely environmental consequences.

The Ministry of Works appealed to the High Court, where the matter was heard by the Chief Justice Sir Ronald Davison. Davison granted the appeal based on a technical argument about the relationship between the Geothermal Energy Act and the Water and Soil Conservation Act, both of which applied to the exploratory drilling.

Not happy with the result, EDS appealed further and in March 1982 the Court of Appeal overturned the High Court and reinstated the Planning Tribunal's decision. The Court of Appeal noted that the tribunal had adopted a test of 'beneficial use' whereby the loss which might follow from the taking of water should be weighed against its benefit. The Court of Appeal thought that such a weighing exercise was a matter of common sense and thoroughly in accord with the purposes of the Act.

As well as stopping exploratory drilling of the Waimangu field, the case established a legal precedent for safeguarding other scenic and natural features in New Zealand. It was to have a profound effect on the country's water law. In an EDS seminar on the future of geothermal energy, Judge Turner explored the then novel possibility of the law actually protecting parts of the natural environment for future generations.

> Are we moving to the point where we say 'there should be no mining in area (a) or (b) or whatever, no matter what

The Waimangu geothermal field in the early 1900s, which developed after the 1886 Tarawera eruption. AUCKLAND LIBRARIES HERITAGE COLLECTIONS 3-ALB63-39

Below: The Waimangu geothermal field in 2023 showing the growth of vegetation around the field. Waimangu is the only geothermal field left in New Zealand that has not been modified by human intervention. RAEWYN PEART

Frying Pan Lake at Waimangu is claimed to be the largest hot spring in the world.
RAEWYN PEART

> the value of the minerals may be'? Are we moving to the point where we should say 'we should not dam this particular river or extract water out of it but leave it in its wild and scenic state and leave it unmodified for our children and grandchildren', without first knowing its energy potential? Are we coming to the point where we can say in relation to one or two geothermal systems, 'well let's preserve it unmodified simply for the sake of preserving it whatever may be the energy potential or whatever else we may derive from it'?

The Ministry of Works' application had been stopped in its tracks, but this was not the only threat to the Waimangu field. While that case was going through the courts, F J Ramsey Limited had built the timber treatment plant. As well as treating wood, the company was planning to produce ethanol out of waste wood, a process that required large amounts of energy. Planning consent for the building was obtained in September 1978 and construction started the following year.

It wasn't until early 1980 that the company obtained licences under the Geothermal Act to operate two bores on the property for production purposes. These were then developed. It was only later that the company made an application for water rights, in October 1982, after the 1981 amendment to the Water and Soil Conservation Act (which clarified that the water rights were needed) came into effect. The application sought to significantly increase the company's take from the two existing bores, from 160 to 900 tonnes per day, and for this to be permitted for at least 10 years.

At the hearing of the application by the regional water board, Ron Keam presented evidence on the uniqueness of the Waimangu thermal area and his belief that the hydrothermal fluid being tapped on the company's property was coming from the Waimangu field. 'The withdrawal is potentially capable of, and may indeed already be, affecting the sensitive and unique cycle of level movement of Inferno Crater Lake,' Keam argued.

A registered engineer, Mr J. H. T. Trust, represented F J Ramsey Limited at the hearing. He gave evidence that, while the applicant had not obtained expert geological opinion, he did not believe that there was sufficient evidence proving a link between the company's bores and the Waimangu geothermal field. However, the company did not produce any scientific evidence to support the application or counter the objections lodged.

After considering the submissions, the regional water board granted consent, but only on very limited terms. The quantity authorised was just 160 tonnes, and the term of the right was just over two years. This was to enable the timber plant to operate in the short term while a potential conversion to natural gas was undertaken. A natural gas pipeline had already been laid on an adjacent farm. However, the company was reluctant to connect to it, as use of geothermal energy was the cheaper option.

Three appeals were lodged against the water board's decision. Those by EDS and the Minister of Lands, who was in charge of managing the Waimangu Scenic Reserve, sought to have the water right refused. The EDS appeal was also in the name of Ronald Keam, Christopher Ecroyd and the Rotorua branch of Forest and Bird. The third appeal, by F J Ramsey Limited, sought to have the terms of the water right amended to provide for a greater take and for a longer term, as sought in its original application.

Geothermal terrace at the Waimangu geothermal area. The geothermal field has remained unmodified due to the efforts of Ron Keam, EDS and others. RAEWYN PEART

The hearing was undertaken over six days in early 1984, after which the tribunal concluded that both the Waimangu and nearby Waiotapu geothermal systems 'have some unique features and should remain undisturbed' and that any long-term right should not be granted. The application was refused and the Waimangu field saved.

The Waimangu thermal area remains unmodified to this day thanks to the efforts of Ron Keam, EDS and others. Much of the awareness raised through their efforts, which highlighted the undisturbed and nationally significant natural and cultural values of the system, has led to it being classified as a 'Protected System' in the Bay of Plenty regional policy statement and regional natural resources plan. Te Mana o Ngāti Rangitihi and the Tūhourangi Tribal Authority currently hold a long-term lease from the Department of Conservation to operate an eco-tourism experience in the area.

When reflecting on EDS's involvement in the geothermal issue, Gary Taylor observed, 'the geothermal case was a precursor to the current debate around limits. Geothermal energy was a resource, that to be used, had to be used within limits. But there was a fairly primitive approach to that issue in those days.'

6. Digging deeper

AT THE TIME THAT EDS WAS FORMED in 1971, the Mining Bill was in the last stages of being passed into law. Submissions to the select committee had already closed and EDS was left with lobbying MPs in Parliament. However, this failed to achieve any changes to the Bill. As reported in the first edition of the *EDS News*, in November 1971, this early experience provided a very important lesson, that 'we have to assert pressure during the formative stages of legislation if we are to achieve any changes'.

THERE WAS A SURGE of mining interest in New Zealand a decade after the Mining Act 1971 came into force and it quickly became evident that the Act had major deficiencies. Gary Taylor summed up the position in the February 1981 edition of *EDS News*: 'in every other environmental controversy there are "rules" that ensure that the worst does not happen, that a reasoned debate and subsequent decision can be arrived at. With mining, though, this is not the case. The "rules" are useless, the Minister decides, and his decision-making is in the absence of any real criteria.'

It was the invasion of Afghanistan by Russia in December 1979, coupled with high oil prices and the Iranian revolution, that helped send the international value of gold soaring in late 1979. The top price, reached in January 1980, was $US850 per ounce. Such high prices prompted mining companies to re-examine old mines and prospects which had earlier been abandoned as uneconomic, as well as to actively seek out new mining areas.

Te Tara-o-te-Ika-a-Māui Coromandel Peninsula was one of the places that caught the eye of local and overseas mining interests. This was because of its multiple deposits of gold- and silver-bearing ore in what had become known as the Hauraki goldfields.

The area had a long history of mining activity after gold was first discovered during the 1860s. Because much of the remaining ore on the peninsula was in low concentrations, opencast mining was likely the most economically viable option for extraction. With the Coromandel's steep slopes and frequent heavy rainfall events, as well as high natural values, such mining risked erosion, contamination of waterways with acid mine drainage and wider ecological damage. Opencast mining would also fundamentally alter the iconic natural landforms.

There was also the issue of the disposal of the enormous quantities of tailings that would be generated from digging large holes in the ground. Such discarded rock, which would be ground to the consistency of fine powder to extract the ore, can contain high levels of heavy metals such as mercury. Safely containing or disposing of such material was a significant challenge.

Early in 1980, EDS got drawn into the mining issue by a Kūaotunu resident who lived just up the road from a potential mining site. As Gary Taylor recalls, 'I was in the EDS office in the town hall one day and Mark Tugendhaft came through the door. He said there was a proposal for this big opencast mine at Otama Beach. I said "you're joking". But he wasn't. So that led to the thick end of 10 years litigating mining cases on the Coromandel Peninsula.'

As Taylor recounted in a speech to the combined Waihi Service Clubs in October 1984, when Mark Tugendhaft first approached him 'he had painted an alarming picture of an opencast mine, of a massive tailings dump right on the beach, and of water and air pollution – and frankly I didn't believe him, I thought he was wildly exaggerating.' But after investigating the proposal Taylor concluded 'he was dead right on every score.'

Previous pages: The steep and rugged Coromandel Peninsula, with its rich deposits of gold- and silver-bearing ore, became a focus for mining interests during the 1980s when the price of precious metals soared. RAEWYN PEART

A New Yorker born in 1949, **Mark Tugendhaft** studied zoology in California before emigrating to New Zealand in 1972 as a 23-year-old keen surfer. He landed a job as a high school science teacher in Gisborne, where the surf was good. Saving his money, with friends he bought a block of land on the Coromandel Peninsula at Kūaotunu. To him it was a 'garden of Eden' located close to a world-class surf break. He built his own house on the land and became a successful potter. When mining was proposed near his home during the early 1980s, Tugendhaft established Peninsula Watchdog with his partner Nedilka Radojkovich to oppose it.

DAVID HARTMAN

Tugendhaft had heard of EDS so he went up to Tāmaki Makaurau Auckland to meet Gary Taylor and persuaded him to become involved in the mining issue on the Coromandel. Tugendhaft worked closely with Taylor and EDS over the following 10 years to successfully head off the mining threat. He eventually turned his potting shed into a coffee-roasting facility and he and Radojkovich established Coffee Lala in 2002.

MARK TUGENDHAFT was outraged that a mining company should seek to destroy his neighbourhood. As he recounted 'a South African company came to the local hall and said "this is what we want to do, we're going to take the whole ridge down to sea level. It will provide more sunlight and 2000 jobs".' Potential job creation was a very salient issue as the Coromandel Peninsula was desperately short of employment opportunities at that time.

Former engineer and EDS director Evan Penny also bought a share in the same Kūaotunu property. He had a slightly different take on the meeting: 'When they first came to the district, they announced they planned to opencast the whole hill down to sea level and fill in the Otama wetland, that "useless" wetland. That was their vision and they thought everyone would jump up and down and think it was great.'

The mining company was Goldmines of New Zealand Limited, a New Zealand subsidiary of the Anglo American Corporation of South Africa. This was the business behemoth that EDS, in support of a small group of 'hippie' residents, looked to take on. The company had identified at least 20 million tonnes of gold-bearing ore on the thumb of land lying between the Whitianga and Whangapoua harbours (colloquially known as the Kūaotunu Peninsula), worth an estimated $1.2 billion. It was seeking three licences to prospect over 2900 hectares, an area which stretched from the coast far back into the hinterland. It included the spectacular Black Jack Scenic Reserve and land directly across the road from where Mark Tugendhaft, Evan Penny and their families had settled.

When Goldmines of New Zealand lodged its applications for prospecting licences, EDS objected, along with many local residents. By now the locals had become organised under the banner of Peninsula Watchdog. Mark Tugendhaft and his partner Nedilka Radojkovich were the prime movers behind the

Goldmines of New Zealand Limited proposed to fill in the extensive wetland behind Otama Beach, shown here, with a giant tailings dam. Such a mine also threatened to pollute the surrounding marine area.
ROB SUISTED/NATURESPIC

organisation, with former journalist Peter Verschaffelt appointed as spokesperson. EDS and Peninsula Watchdog worked closely together, EDS providing the legal nous and Peninsula Watchdog mobilising people on the ground and raising funds for the campaign.

When EDS first became involved in the mining issue, as Taylor explained, the companies treated the society with

> contemptuous disregard and seemed determined to ignore us. Then we took them to court and they started to take notice. They realised that they were not going to have things entirely their own way on the Coromandel Peninsula and their tactics changed. Instead of extolling the virtues of their mining operations and perhaps exaggerating the scale and benefits, they began to be much more restrained in their public comments.

The hearing of objections to the Goldmines prospecting licences commenced on Monday 16 February 1981 in the Thames District Court.

The presiding judge, John Kenneth Patterson, was a former mines inspector who had grown up in a mining town on the West Coast of the South Island. Tony Randerson, then a partner in law firm Wallace McLean Bawden, was leading the EDS case.

A number of witnesses gave evidence, which mainly related to the potential effects of opencast mining on the Kūaotunu Peninsula and surrounding marine area. The coastline was rich in shellfish, including scallops, pīpī and cockles. Such filter feeders would be particularly susceptible to the fine sediment run-off that opencast mining would likely generate.

On the Wednesday of the hearing, Judge Patterson announced that he would no longer hear evidence about the impacts of mining and any earlier such evidence would be 'struck from the record'. This was on the basis that it was not relevant to the application, which was only for prospecting not mining. The reason EDS and others were presenting evidence on mining impacts was that this would be the only opportunity to do so. At that time, under the Mining Act, once a prospecting licence had been issued it provided an automatic right to a mining licence without any further public input.

After Judge Patterson decided to exclude evidence related to mining impacts, the EDS team held an urgent meeting to consider potential next steps. The ruling effectively excluded most of their case. A decision was made to seek a High Court review of Patterson's decision. Tony Randerson raced back to Auckland and spent the night drafting papers seeking an interim injunction. Arranging for a law-firm staffer to urgently file the papers in the Auckland High Court, Randerson drove back down to Thames the next morning.

By 12.20pm on the Thursday, EDS had a decision from the High Court, and it was favourable. Justice Holland found that there was a substantive case to be argued. He did not order the District Court hearing to be stopped, but he prohibited Judge Patterson from concluding his investigation until the substance of the challenge had been determined by the High Court.

While the interim injunction was being determined in Auckland, Randerson was back in Thames, appearing in front of Patterson. When he heard about the successful interim injunction, Randerson advised the court. However, as he recalls,

> the judge refused to stop the hearing until he saw a sealed copy of the order. I said it would be there around 4 o'clock. At precisely that time the doors to the courtroom swung open and Gary [Taylor] came in holding a piece of paper. It was the sealed order. I handed it to the judge who was pretty confused by then. The following day he stopped the hearing.

On that last day of the hearing, Gary Taylor had joined Energy Minister Bill Birch on a helicopter inspection of the land affected by the prospecting application. They were accompanied by a geologist working for Goldmines (Dr Franco Pirajno) and an official from the Mines Division. The party flew over the Kūaotunu Peninsula, which looked especially impressive from the air, and landed on Bald Spur, a prominent ridge within the application area. The party then inspected the farm running behind Otama Beach, which was the likely location of the tailings dam in the event that opencast mining took place.

Taylor recounted what followed in the May 1981 edition of *EDS News*.

I told the Minister that EDS geologist Simon Carryer had calculated that tailings from an opencast mine based on an ore body of 30,000,000 tonnes [which was the company's estimate of size] would fill this basin to a depth of 30 metres. Mr Birch was sceptical and said he doubted that that could be the scale of things. Dr Pirajno proceeded to do his own calculations on the spot – and quickly confirmed that our figures were 'about right'.

The incident is worth recounting, because it demonstrates that Mr Birch – and other pro-mining exponents – did not fully appreciate that 1980s-style mining is fundamentally different to what took place previously. The scale of mining proposed today is enormous. For the first time, perhaps, the Minister of Energy could visualise the stark reality of what could happen. It is not emotive or an exaggeration, or distortion – it is fact.

The EDS case to save the Kūaotunu Peninsula (including Otama Beach shown here) from an enormous opencast goldmine resulted in changes to the Mining Act 1971 so that a prospecting licence could no longer automatically be converted into a mining licence. RAEWYN PEART

The substantive challenge against Judge Patterson's ruling was heard in the High Court during late June and early July. In the first instance, the Crown's lawyers challenged EDS's standing to bring the proceedings at all. This was quickly dismissed by Justice Speight on the basis that 'the Society was a reputable group of Auckland citizens who clearly had an interest in protecting the environment of the Coromandel Peninsula and thus had standing.'

Justice Speight's final decision was released on 10 July and it largely agreed with the EDS case. The judge confirmed that as there was no public objection right to mining licences, the prospecting stage was the only opportunity for the minister to be apprised of information that objectors might raise before granting the right to mine. He ordered Judge Patterson to consider evidence which was realistic as to the nature and consequences of future mining in the areas covered by the prospecting applications. Costs were awarded in favour of EDS.

As author Roger Wilson observed, 'this

landmark decision was clearly influential in having the "automatic right to mine" in the Act removed, since it highlighted the impossibility of conducting a serious argument on the merits of mining an area which had yet to be prospected.'

The prospecting licence hearing reconvened in early October 1982, almost two years after it had first commenced. By the end, EDS had called 11 witnesses to give evidence on the environmental impact of large-scale mining. One of them was Auckland University Professor John Morton, who testified on the value of the Kūaotunu coastline and coastal waters. Graeme Lawrence, who at that time was the Thames-Coromandel District Planning Officer, recalls,

> at one stage during the hearing Judge Patterson said to Professor Morton 'you obviously know your subject. But if you were walking along the coast at Otama, and you looked down and saw gold shining in the sand, what would you do?'. Morton said without batting an eyelid, 'Your Honour, I hope I would resist the temptation to bend down and pick it up.'

COURTESY OF TONY RANDERSON

Hon Tony Randerson CNZM KC graduated from the University of Auckland with a Bachelor of Laws degree in 1969. In 1972 he became a partner in the law firm Wallace McLean, which later merged to become part of Kensington Swan. In 1989 Randerson left the firm to operate as a barrister sole. The following year he chaired the review group that finalised the Resource Management Act 1991. He was appointed a Queen's Counsel in 1996.

Randerson was appointed to the High Court in 1997 and became Chief High Court Judge in 2004. He was subsequently appointed to the Court of Appeal in 2010 and retired from the bench in 2017. In 2019, Randerson chaired the Resource Management Review Panel that recommended the Resource Management Act be replaced with new legislation, including a Natural and Built Environment Act. In 2021, he was appointed a Companion of the New Zealand Order of Merit for services to the judiciary.

Randerson was a director of EDS from 1985 to 1987 and he ran several cases for the society, including the Otama opencast mine case.

Fishing Industries Board representative Alec Duncan and local fishermen also supported the EDS case on the basis of potential impacts on the fishing industry if pollution occurred. EDS subpoenaed witnesses from the Thames-Coromandel District Council, who supported placing a moratorium on mining on the Coromandel Peninsula until a proper planning study could be carried out. In addition, the Commission for the Environment questioned the wisdom of proceeding with a massive industrial complex in a rural recreational area.

The mining company only called two witnesses, one who was a company geologist. It gave no evidence disputing the EDS witnesses' contentions, claiming that much of it was irrelevant. The company later said it would 'do the right thing' and it had the resources to plan the tailings dam properly to meet all environmental objections. It could not say what the mine or mines would look like, until prospecting revealed the true resource, but admitted that any mine would be a sizeable one.

During the course of the hearing, Judge Patterson went out into the field to view the area covered by the prospecting applications. As Tony Randerson recalls,

> we took a helicopter ride from the Thames airfield right around the entire Coromandel Peninsula and then also to the Kūaotunu Peninsula where Otama and Opito are. It was just a stunning coastline when you see it from the air. But the judge was from the West Coast and he thought mining would be the best thing for Thames. He told me during the helicopter trip, 'Thames would boom.'

Mark Tugendhaft recalls Patterson telling him, 'I grew up in a mining town on the West Coast and I really enjoyed my childhood, so I'm not sure what you're worried about.' Gary Taylor found Patterson to be a courteous man who conducted the hearing in a fair and professional manner. 'But I thought he was out of his time. He got overtaken by this wave of opposition from people who were much younger than him and did not want the kind of world he was there to deliberate on.'

In the end, Judge Patterson recommended that two of the prospecting licences be issued, and the third at Opito be refused. The applications were then sent to the energy minister for a final decision.

WHILE THE KŪAOTUNU mining proposal was being dragged through the courts, a head of steam was developing behind the need to fix up the Mining Act itself. The legislation had been developed in an era when only small-scale mining activities were contemplated. It was clearly not up to the task of dealing with the avalanche of applications now driven by the high price of precious metals. Some of the worst aspects of the Act included the automatic right to a mining licence after a prospecting licence had been granted, the ability to mine without landowner consent, and the lack of proper environmental scrutiny of mining activities.

In November 1980, EDS lodged its first submission with Energy Minister Bill Birch seeking to have the Mining Act updated. In particular, the society wanted mining activities to have the same level of scrutiny as other activities which came under the Town and Country Planning Act.

At the same time, Peninsula Watchdog instigated a series of public meetings to demonstrate the strength of public opposition to large-scale mining on the Coromandel Peninsula. Between September 1980 and February 1981 meetings were held in Colville, Te Puru, Coromandel, Whitianga, Coroglen, Whangamatā, Kūaotunu, Katikati and Thames. In addition, a petition was circulated calling for the reform of the Mining Act.

In response to the upwelling of public concern, Minister Birch visited the Coromandel Peninsula in February 1981, and while there he made a commitment to amend the Mining Act. Soon afterwards, newly appointed Environment Minister Dr Ian Shearer also toured the peninsula. He came away deeply concerned about the potential effects of large-scale mining on the area.

Other politicians were more gung-ho. The Hauraki Member of Parliament, Leo Schultz, proposed using mine tailings to construct a causeway all the way from the Coromandel Peninsula to Auckland. He also suggested using tailings to block each end of the Karangahake Gorge in order to form a recreational lake. There were also suggestions that mine waste could simply be dumped into the Hauraki Gulf.

Minister Birch commissioned Link Consultants to undertake a review of the Act and gauge public opinion as to its efficacy. The report emerged in late March but did not go as far as EDS wanted. In May 1981, the society, along with Forest and Bird, lodged a further extensive written submission to Minister Birch. At this point the government moved quickly. The Mining Amendment Bill was introduced into Parliament on 16 July and was passed just three months later on 23 October. It came into legal effect on 1 January 1982.

The changes to the original Act included the deletion of the automatic right to exchange a prospecting licence for a mining licence and greater opportunity for public and local authority participation. In addition, objections were now to be heard by the specialist Planning Tribunal rather than a District Court judge.

As Tony Randerson later observed, 'this represented a major step forward. The adverse environmental effects of mining operations could be properly investigated and determined by an independent body with appropriate expertise to assess such effects.'

The Amendment Act reflected EDS's input to a significant extent. The society had made submissions to the select committee and had followed these up with supplementary submissions sent direct to the Minister 'at the eleventh hour'. As Gary Taylor explained,

> the changes to law can be criticised for some omissions and irregularities but in general must be seen as a significant victory for the environmental movement. Very strong lobbying by the mining industry failed to gain any important concessions. The new procedures lift the Mining Act out of the dark ages and into a system of proper evaluation of the environmental impacts of mining.

Later in 1982, EDS published *An Objector's Guide to the Mining Act 1971* designed to ensure that members of the public could make maximum use of the opportunities provided by the new provisions in the Act.

DESPITE THE AMENDMENTS to the Mining Act, which meant that new licence applications would be put under greater scrutiny, there were still concerns that the situation on the Coromandel Peninsula was out of control. In the six-year period up until 1982, the Thames-Coromandel District Council had considered some 100 applications related to mining. Over a quarter of the peninsula was affected by mining applications of one kind or another. There were concerns that each application was being considered in a piecemeal way without consideration of the broader context. The council was keen to undertake a broader study focused on the planning implications of prospecting and mining on the peninsula, and in the meantime, it wanted a moratorium on the granting of any more prospecting or mining licences.

EDS was right behind the council, and its lawyers drafted the Thames-Coromandel District Council Mining Moratorium Bill. This was designed to place an 18-month moratorium on the issue of licences while the planning study was completed. The council gave the Bill the go-ahead and then it was a matter of getting it introduced into the House. After a short delay, during which local MP Graeme Lee

was accused of stalling, he finally introduced the Bill on 29 April 1983 and it was sent to the Local Bills Select Committee for scrutiny.

Meanwhile, work on the planning study commenced. It was led by David Burton, a planner from the Ministry of Works and Development. The Bill never re-emerged but the planning study did, in mid-1984. It largely took a compromise approach, suggesting that some land be made available for mining through a zoning approach. This was not what many residents wanted to hear.

The northern west coast of the Coromandel Peninsula. EDS helped the Thames-Coromandel District Council draft a mining moratorium bill to stop the issue of mining licences on the peninsula while a planning study was completed. The Bill was introduced into Parliament but never proceeded. RAEWYN PEART

PRIME MINISTER Robert Muldoon called a snap general election in July 1984. The Labour Party won and new Energy Minister Bob Tizard announced that the government would not permit large-scale opencast mining on the Coromandel Peninsula. By this time, EDS had been fighting Goldmines of New Zealand in the courts for five long years. But the society was now running out of legal runway. In December 1984, and despite the earlier announcement, Minister Tizard decided to grant all three prospecting licences to the company.

Horrified, Gary Taylor and Mark Tugendhaft mounted a flurry of intensive lobbying of the new

Otama Beach at sunset. EDS and Peninsula Watchdog fought Goldmines of New Zealand Limited for years through the courts to stop mining at Otama and elsewhere on the Kūaotunu Peninsula. The company eventually obtained three prospecting licences but was unable to undertake prospecting work on the ground due to direct action by Watchdog supporters. ROB SUISTED/NATURESPIC

government ministers. They met with Deputy Prime Minister and Attorney-General Geoffrey Palmer, Environment Minister Russell Marshall and Minister of National Development David Caygill, as well as Prime Minister David Lange. This prompted the issue to be discussed at the last caucus meeting of 1984, after which the Prime Minister announced there would be a freeze on the Goldmines application pending review. But this provided only a short-term respite.

In a last-ditch attempt to derail the process, EDS filed judicial review proceedings in the High Court. These alleged that the energy minister had not consulted with his colleague the Minister of Lands before granting consent as required under the Act. Records which EDS had obtained from the Mines Division indicated that such consultation had only taken place after the decision had been made. However, they did not prove to be correct. Once the proceedings were filed, the Minister of Energy produced an affidavit swearing that he had in fact undertaken the requisite consultation. At that point EDS's case collapsed and the society was forced to withdraw proceedings in order to avoid a costs award. It was now July 1985.

At that point Goldmines successfully acquired its three prospecting licences and its workforce attempted to go on site to undertake the work. But by that time Peninsula Watchdog had established a formidable fighting force on the ground. Members engaged in direct action when any miners appeared. They mounted roadblocks and sit-ins. They cut vehicle tyres.

And they concreted up workers' toilet facilities and put dead possums in their water supplies. As Mark Tugendhaft recalls,

> someone called up and said 'the prospectors are in town'. So we went up to the gate of the property and when the truck drove up we stood in front. While they were there talking, Nedilka went into the car, took out the keys, and threw them into the bush. When they headed into town to get a new set of keys someone cut their tyres. We just stuffed them every time they came, day after day.

In the face of such staunch opposition the prospectors eventually gave up trying to access the Kūaotunu land. By August 1985, Goldmines had announced its intention to withdraw operations from New Zealand entirely. The company was deregistered in 1992.

AS THE OPPOSITION to opencast mining grew, the industry started to focus more on underground mines. One of these was proposed at Monowai, some 13 kilometres north of Thames, and within the Coromandel Forest Park. An associated processing plant was to be located at Maratoto, some 60 kilometres south. A New Zealand-owned company, Spectrum Resources (formerly Crusader Minerals and later Charlie's Group Limited), applied for a mining licence for the site in October 1983. It was the second mining licence to be considered after the amendments strengthening the Mining Act had come into force.

Peninsula Watchdog (now formally incorporated as Coromandel Peninsula Watchdog) and local residents, under the banner of the Waiomu Action Group, were very concerned about the proposal. They contacted EDS. Under the new procedures, Spectrum Resources was required to prepare an environmental impact report, which it did. But on reviewing the report EDS found it to be grossly defective. In particular, it failed to describe where or how the tailings would be disposed of beyond the first four years of operation. It looked like the company was endeavouring to get away with doing as little work as possible for its application, effectively trying on the system.

EDS lodged proceedings in the High Court seeking a declaration that the deficiencies in the report were so large that it was, in effect, a nullity. The Commission for the Environment, which was charged with auditing the report, subsequently announced that it also considered the report to be deficient. The commission prepared a further 173 questions to which the company was required to respond. Spectrum Resources then did more work, substantially modified its proposals, and produced a supplementary report.

As well as a mining licence, the company required water rights and EDS objected to these as well. A key concern for the society was the impact of polluted water draining from the mine. Spectrum Resources proposed to simply plug the mine after the ore had been extracted. But EDS thought this was grossly inadequate. As explained in the December 1986 issue of *EDS News*, 'they intend to simply put a concrete wall in the front of the mine and assume that that will be a permanent barrier to leachate from the mine reaching the adjacent stream. We do not accept that this is a professional or practical method of dealing with long-term waste from the mining operation.'

A further concern related to the processing

Indigenous forest in the Waiomu Valley. Once opposition to opencast mining on the peninsula grew, the mining industry started focusing on underground mines and in particular the Monowai mine in the Waiomu Valley of the Coromandel Forest Park. RAEWYN PEART

plant where the company proposed to establish a tailings disposal area. EDS was not convinced that the plan to seal the ground beneath the tailings was adequate. The society was concerned that the waste effluent, with all its heavy-metal contaminants, could percolate down into the groundwater. In the end, the Hauraki Regional Water Board was persuaded of the risks and declined to grant three of the water rights. The company subsequently appealed.

As the mine was within the Coromandel State Forest Park it required consent from the Minister of Forests. At the time, the minister was Koro Wētere and he gave consent in March 1986. In September that year, Energy Minister Tizard published proposed conditions for the mining consent which were open to public objection. EDS lodged an objection along with others and these were passed on to the Planning Tribunal for hearing under the new Mining Act provisions. The Thames-Coromandel District Council was among the objectors. The council was seeking to protect the quality of private water supplies on which the coastal settlement of Waiomu relied.

Before the matter could be heard by the tribunal, the Conservation Act came into force on 1 April 1987, and the forest land in question became part of the conservation estate managed by the new Department of Conservation. Helen Clark was appointed the new Minister of Conservation.

The case was set down to be heard by the Planning Tribunal on 7 September 1987. EDS found the proposed conditions to be grossly inadequate, and so it launched proceedings in the High Court, seeking to have the Planning Tribunal hearing delayed while the conditions were rewritten and readvertised. Tony Randerson appeared for EDS and the Waiomu Action Group.

The High Court hearing was held on 30 July 1987. Justice Richard Heron was sympathetic to EDS's assertions that the proposed conditions were 'perfunctory and somewhat rudimentary in their application to this application overall' and that they conferred on the Inspector of Mines discretions and authorities not authorised by the Mining Act. In effect, the Minister was delegating the role of regulating the mine to the inspector. However, the judge did not consider the deficiencies to be so great as to prevent the Planning Tribunal from embarking on a hearing. He concluded that the tribunal could amend the conditions to fix the problems after hearing the submissions of the parties.

The Planning Tribunal hearing went ahead in Thames as scheduled, with Judge David Sheppard presiding. Coromandel Peninsula Watchdog sought an adjournment of the inquiry on the basis that the new Minister of Conservation had not given consent to mining on what was now conservation land. But this was dismissed by Judge Sheppard on the basis that the earlier consent provided by Wētere had not been revoked, so was still extant.

The hearing itself took three months. EDS was unable to maintain a presence over such a long period and the heavy lifting was done by former EDS director Evan Penny, who represented Coromandel Peninsula Watchdog, and local resident Margaret Pye who represented the Waiomu Action Group. Geologist Simon Carryer gave evidence to support the EDS case and managed to expose a gaping hole in the application. Spectrum Resources could not guarantee management of acid drainage from the underground mine as the proposal went outside the scope of the geological information it had for the area.

Judge Sheppard concluded the hearing on 1 December 1987. But the Planning Tribunal never got to release its recommendations. Later that month, on 22 December, Conservation Minister Helen Clark revoked consent to undertake mining on forest park land. Spectrum Resources judicially reviewed the ministerial decision in the High Court and succeeded in overturning it. But then Helen Clark reconsidered the matter and declined consent again. That was the end of the Monowai mine.

Kūaotunu and Waiomu were not the only places under threat from miners, as there were applications over a large swathe of the Coromandel Peninsula, as well as elsewhere in New Zealand. EDS was involved in opposing many other mining proposals through regional water boards, the Planning Tribunal and the higher courts, but largely without success. The law was oriented too strongly in favour of the miners. But the legal proceedings delayed things and gradually wore the mining proponents down. In the end, no large-scale opencast mining took place on the Coromandel

Peninsula proper, but only further south at Waihi (Martha Mine) and partly at Waitekauri (Golden Cross Mine), where a portion of the mine was also underground.

EDS and Peninsula Watchdog may have lost some battles, but by joining their respective forces they eventually won the war. Or at least so far.

—000—

ONE OF THE LAST mining cases EDS was involved in was at Waihi. The application for a mining licence to establish a fresh opencast mine on the old mine-workings at Martha Hill was lodged in April 1985 by the Waihi Gold Company. The proposal involved removing some 33 million tonnes of rock and other material from Martha Hill, which was then slated to become a lake once the mine was exhausted.

The reason EDS got involved, as Gary Taylor explains, was that

> the community was under siege and they needed help. These were lay people who had houses above the tunnels that made up the Waihi mine. We were concerned about what was a very large tailings dam and the need to ensure that it was properly designed and reinforced against earthquakes and the like. And the Waihi mine was to be a big ugly hole in the middle of the town which was not without its own environmental impacts. But the main reason we got involved was the social connections that EDS had built up with the community and the expectation that we would help them out with the more technical hearings process.

In the end, the society represented 86 objectors, an enormous number to manage. The legal team consisted of Auckland barristers Raynor Asher and Paul Cavanagh. There were two of them so that they could split up the considerable case load. Gary Taylor coordinated the case.

The hearing ran for 26 days and evidence was given by 53 witnesses, many who were experts. EDS was present during the entire hearing, cross-examining the applicant's witnesses, and presenting its own, including a noise consultant, social-impact assessors and a freshwater biologist. The society was *the* major environmental advocate at the hearing with no government agency evidencing concern for the protection of the environment.

Gary Taylor recalls one particularly notable incident during the long hearing,

> during a morning session we had been hearing evidence about the seismic integrity of the proposed tailings dam. Over the lunch break I was sitting in the hearings room preparing material for the afternoon session and Judge Sheppard was working at his bench. I recall we were the only people in the room. With some kind of ironic serendipity, a significant earthquake occurred, rattling the bones of the old weatherboard building and causing the Judge and I to look at each other with eyebrows raised. This was before the days of 'drop, cover and hold'. The hearing proceeded after lunch and nothing more was said about the earthquake.

In the end, the tribunal granted consent but subject to much tighter and more enforceable conditions than had been originally proposed.

EDS became involved in the Waihi opencast mining case in order to support the local community in what was a highly complex and technical case. The society represented 86 objectors and maintained a continual presence at the hearing which ran for 26 days. The early development of the mine is shown here.
LLOYD HOMER, GNS SCIENCE

This was largely due to EDS's involvement. The hearing considerably stretched EDS's resources. As described in the July 1987 edition of *EDS News*,

> The Waihi hearing was basically too big for the Society to handle and we were faced with huge costs far beyond our budget – although of course our budget was minute compared with what the applicants would have invested in the hearing. While it might be possible for us to ask for free legal advocacy from lawyers for a hearing lasting perhaps a week, a five-week hearing is far too long, and the fact that multiple advocates were required, detracted from our effectiveness.

When the Resource Management Act came into force in 1991, mining was squarely included within the Act's provisions, and required to undergo the same environmental scrutiny as other activities. This was something EDS had been advocating for over many years. A separate allocation regime for minerals was provided for in the Crown Minerals Act 1991.

Six years later, in 1997, the Crown Minerals Act was amended to prevent mining north of the Kopu-Hikuai Road on the Coromandel Peninsula, as well as on most conservation land and offshore islands in the Hauraki Gulf, including Aotea Great Barrier Island. This was largely a result of the continuing efforts of Coromandel Peninsula Watchdog, although as Mark Tugendhaft observed, 'getting the amendment passed wouldn't have happened without EDS'.

Like many environmental issues, there is likely to be more to play out with this one.

7. Protecting the coast

FROM ITS INCEPTION, EDS sought to protect the high natural values of New Zealand's impressive coastline. An area of particular focus was the Coromandel Peninsula, which was coming under increasing urban development pressure. EDS's work on the peninsula was led by Laurie Newhook. As a young lawyer, Newhook cut his teeth on EDS cases. He would later go on to become the Chief Judge of the Environment Court.

EDS'S FIRST COASTAL skirmish was over Waikawau Bay, a remote undeveloped beach on the north-eastern coast of the Coromandel Peninsula. The bay has a long, gently curving white-sand beach, two estuaries and an extensive wetland system. The land behind the southern end of the beach was owned and farmed by Andrew and Mavis Goudie, who had lived there since the 1930s.

In September 1971 Advantage Investments Limited, a Hamilton-based syndicate headed by bank manager C. G. Grice, announced plans to develop the beachfront and subdivide the hills behind the beach to create up to 1500 low-intensity residential sections. The development was ambitious and included an airfield running alongside the estuary and down to the beach; a private golf course immediately behind the dunes; camping grounds and caravan parks at both ends of the beach; a central complex containing a restaurant, bar, gymnasium, swimming pool and squash and tennis courts; a commercial area for shopping; and eventually a boat marina.

There was also to be 'extensive planting of native and exotic trees to give each section a truly rural setting'. In perhaps one of the earliest claims that such developments would not detract from, but indeed only serve to improve, the coastline, insurance broker and syndicate member F. G. Baker was reported as stating 'We want to leave the land as it is – in fact in a very real way we will be restoring it to what it was.'

Previous pages: Waikawau Bay, Coromandel Peninsula. In the early 1970s, EDS and others managed to head off a proposal to develop the land behind Waikawau beach into a large resort area with up to 1500 houses, recreational facilities, restaurants and shops, a golf course and an airstrip. RAEWYN PEART

Advantage Investments had reportedly reached agreement to purchase not only the Goudie beachfront farm but the one to the north owned by the McNeils. This farm included the northern headland overlooking the beach. The company was seeking to change the planning rules, including rezoning both the northern and southern headlands for subdivision.

Even though the farmers were ready to cash up, there was one landowner in the bay who wasn't. M. Andrews was custodian of a small strip of Māori-owned land that ran up from the shoreline into the middle of the beachfront land. She refused to sell, stating that she wanted to make the land available

Surfer at Waikawau Bay. After the development proposal for Waikawau beach was headed off, the government purchased part of the Goudie farm and leased an additional area of land. This is now collectively managed as a recreational reserve and accommodates the Department of Conservation's largest campsite. RAEWYN PEART

for members of the public to camp on.

EDS met with the developers to find out what was proposed and then lodged a strong objection to the development. The society also provided assistance to others who were opposed. Conservationist Dr Ron Locker was particularly vocal in opposing the scheme as he had been pursuing the idea of setting aside a network of coastal reserves on the Coromandel Peninsula. Evidently put off by the level of opposition, which was unusual at the time, the company abandoned its development plans and the Goudie family continued to farm the land.

EDS was keen to get the farm into public ownership before further development proposals came along. Cabinet considered a proposal to buy the land on several occasions but deferred a decision due to lack of money. There was also a dispute over the value of the property, with the Goudies asking for double the government's initial valuation. Finally, in 1976, the government purchased part of the Goudie farm, which was then designated as

Onemana under construction in 1974. Onemana, known to Māori as Whitipirorua, was developed over an early Māori settlement supported by abundant seafood and obsidian and chert for tool-making. WHITES AVIATION, ALEXANDER TURNBULL LIBRARY

a recreation reserve. Further land behind the beach was leased from the Goudie family in 1980 in order to extend the reserve. It is now managed by the Department of Conservation and is the department's largest campsite, accommodating up to 660 people. This was a major achievement, but still left the northern headland in private ownership; an issue that EDS was to return to.

—000—

THE WAIKAWAU DEVELOPMENT proposal was soon followed by another. Just 9 kilometres north of Whangamatā lies a stunning white-sand surf beach bounded by rocky cliffs to the south. Known to Māori as Whitipirorua, the area had featured a substantial early Māori kāinga supported by abundant seafood and locally available obsidian and chert for tool-making. Ngāti Hei was the earliest known tribal presence in the area but had been supplanted by Ngāti Hako in the mid-seventeenth century.

During the early 1970s, landowner Brian Bambury planned to subdivide the rolling hills behind the beach into 364 urban lots. EDS lodged an objection to the proposal, but the Thames County Council approved it anyway. The society then appealed to the Town and Country Appeal Board. Laurie Newhook appeared for EDS at the hearing in Thames, which was chaired by Judge Arnold Turner. It was April 1973 and Newhook was just 22 years old.

Brian Bambury challenged the ability of EDS

Onemana Beach. Earthworks for the development of Onemana during the 1970s destroyed part of the early Māori settlement site; a tennis court and toilet block (which can be seen on the reserve in front of the houses) were built over its remains. EDS tried to stop the development but was refused standing in front of the Town and Country Appeal Board to do so. RAEWYN PEART

to bring the appeal on the basis of standing. At that time, the Town and Country Planning Act provided an appeal right only to parties that were 'affected' by the proposal. EDS was not directly affected but was taking the case in the public interest. The Appeal Board held that this was not sufficient and EDS's appeal was dismissed without the opportunity to present evidence or submissions. The proposal for the development was then granted consent.

By 1974 the subdivision, called Onemana, was underway. Earthworks destroyed part of the early Māori settlement site and a tennis court and toilet block were built over its remains. Sadly, EDS had been unable to protect the natural and cultural values of this part of the coastline due to a lack of standing in the law.

IN 1972, THE COROMANDEL County Council publicly notified its district scheme and EDS put it under close scrutiny. It found that the scheme had been over-generous in providing for coastal subdivision and contained few provisions to protect the natural values of the coastline. As reprised in the September 1976

Matarangi Peninsula in 1976, when there was still little development on the sandspit, although a subdivision had been created at the eastern end right on the frontal dunes.
WHITES AVIATION, ALEXANDER TURNBULL LIBRARY

edition of *EDS News*, the district scheme as notified in 1972 'would have allowed almost uncontrolled development throughout the peninsula'.

Along with others, EDS lodged extensive submissions arguing for greater protection of the coast. In response to the numerous objections lodged, the council decided to pause the district scheme process in order to undertake further work on its approach to coastal management.

The Town and Country Planning Amendment Act 1973 was passed in November the following year. It inserted a new section into the Act requiring councils to address a number of matters of national importance in their district schemes. One of these was 'the preservation of the natural character of the coastal environment and of the margins of lakes and rivers and the protection of them from unnecessary subdivision and development'. The amendment had arisen directly out of growing concerns about the impact of poorly managed developments around the country's coasts, including on the Coromandel Peninsula.

Following the legislative change in October 1974, the Coromandel County Council publicly notified Variation 1 to its proposed district scheme. This provided for an 800-metre-wide coastal zone around most of the Coromandel coastline outside urban areas. This spatially identified a sensitive coastal strip which would be subject to more restrictive planning controls.

EDS supported the general approach but thought the rules applying to development within the zone were too weak. In particular, subdivisions could still be permitted almost anywhere within it as a conditional use (analogous to a discretionary activity).

EDS made submissions to strengthen the provisions and ended up in front of the Town and Country Appeal Board in August 1976. Laurie Newhook represented EDS, as well

Houses at Matarangi in 2023. Although the considerable erosion risk affecting the early subdivision was identified during the 1970s, when only seven houses had been built, the area has continued to be developed and many houses there are currently threatened by coastal erosion. RAEWYN PEART

as the Physical Environment Association of Coromandel, a local group established in 1972 to help protect the natural environment of the peninsula. David Dodds gave expert planning evidence in support.

The matter was eventually settled through the parties agreeing to substantially strengthen the planning controls. The provision for conditional uses within the coastal zone was deleted. This meant that residential development could only take place there if the developer established a need for it and it could meet other plan requirements to protect the natural environment. In effect, the coastal zone had become a conservation zone.

The Coromandel coastal zone not only strengthened protection of the Coromandel Peninsula's coastline, but it also provided a useful model that was adopted in other parts of the country, including Northland.

IN 1975, THE COROMANDEL County Council was amalgamated with Thames County and Thames Borough to become the Thames-Coromandel District Council. That same year, the council notified Variation 7 to its district scheme to enable further development of the Matarangi Peninsula. EDS assisted the Physical Environment Association of Coromandel to object to the variation.

The Matarangi Peninsula is a large, low-lying sandspit located some 15 kilometres north of Whitianga. It is situated on the north-west flanks of the Kūaotunu Peninsula and extends across the mouth of the Whangapoua Harbour. The Coromandel County Council had granted approval in 1968 to subdivide the eastern end of the peninsula into 115 lots. Prior to EDS becoming involved in the case, the council had subsequently included a 'Future Development Zone' in its district scheme, indicating that much of the remaining sandspit could also be developed. But before such development could occur, a structure plan needed to be incorporated into the district scheme to provide more detailed design and controls. This was the purpose of Variation 7.

The design of the initial subdivision had been blind to the risk of coastal erosion, with the frontal sand dunes flattened to provide building platforms, and house sites located little more than 20 metres from the high-tide mark.

The development of the subdivision had proceeded slowly, and by the time the district scheme variation was notified in 1975, only 13 lots had been sold and seven built on. Despite this, the developer was keen to secure further subdivision rights on the peninsula. The proposed structure plan included two 'residential cottage cluster' areas close to the ocean beach, a 'tourist facilities' zone to provide for motels, hotels and restaurants further to the west near the tip of the sandspit, and additional commercial and residential zones.

EDS arranged for Laurie Newhook to appear in front of the Town and Country Planning Appeal Board on behalf of the Physical Environment Association. Newhook was up against Tāmaki Makaurau Auckland barrister Peter Salmon, who was appearing for the council, and lawyer David Sheppard, who was representing the developer, Kenwood Properties.

To give evidence in support of the structure plan, Salmon called a council-commissioned consultant planner as well as a planning officer employed by the Ministry of Works and Development. Expert witnesses for Kenwood Properties gave further evidence in support of the proposal, including on the provision of water supply and sewage services to the development.

Newhook had no expert witnesses at his disposal to directly challenge the evidence presented by such heavyweights. There had been no funding to engage such technical support. However, he was able to present two lay witnesses. The first was Sir Robert Falla, a former Chairman of the Nature Conservation Council, who spoke of the natural values of the peninsula and it being one of the few remaining natural spit features on the east coast of the North Island.

The second witness, Harold Brown, had previously been a pro-development councillor, but was now a member of the Physical Environment Association. He spoke strongly about the risk of coastal erosion on the sandspit, the difficulties of providing water and sewage services, and the problem of upgrading the road.

Although not technical in nature, this evidence was sufficient to identify a gaping hole in the proponents' case. It demonstrated that neither the council nor the developer had seriously grappled with the engineering issues associated with building on a sandspit, including how to manage erosion risk. The appeal board was clearly unsettled by this and decided to call for more evidence on the matter. The hearing was adjourned for almost a year while the opinion of the Hauraki Catchment Board's chief engineer was obtained.

This advice, when it became available, recommended a total ban on structures within 100 metres of the seaward toe of the frontal dune and that only relocatable structures should be permitted within the next 100-metre buffer strip. The engineer also advised against developing within 120 metres of high tide on

Due to the efforts of EDS and the Physical Environment Association of Coromandel the second stage of subdivision at Matarangi was set well back from the beach and coastal erosion risk. RAEWYN PEART

the harbour side of the spit and within 2800 metres of its tip. He noted that the existing approved development was almost wholly within the area where he advised structures should be prohibited.

The implications of applying such buffer zones to the structure plan put forward by the council were profound. The proposed 'tourist facilities' zone would need to be eliminated entirely and portions of the 'residential cottage cluster' zones and 'service commercial' zone would need to be removed.

It was clear that the design of the development needed to be substantially changed. When this became evident, the appeal board effectively threw out the variation, and instructed the council to go back to the drawing board. With few resources, and against all odds, Newhook had managed to win the case.

As explained in the November 1977 edition of *EDS News*, the case had achieved a number of things. It had bought time before the development could occur and much more careful planning would be required if development on the peninsula was to be proposed in the future. In addition, it also demonstrated to council that 'those who care for the natural environment of the Coromandel Peninsula will not sit quietly by and witness unacceptable development' on it.

The Matarangi Peninsula was subsequently developed under a revised structure plan. The houses are set well back from the coast. The 1977 appeal board decision was fortuitous, as the houses built as part of the first subdivision are now facing serious erosion and flooding risks. If EDS had not become involved in support of a local community group, it seems

almost certain that the second stage of the subdivision would also now be similarly threatened.

Potentially millions of dollars have been saved in relocation and protection costs as a result of intervention by the Physical Environment Association and EDS. The mobile outer end of the sandspit, with its higher value natural landscape and ecosystem qualities, has also consequently remained relatively undeveloped.

—ooo—

BY THE LATE 1970S, development pressure on the Coromandel Peninsula had become relentless. The next focus for EDS was Hahei, located on the east coast of the Coromandel Peninsula just south of Whitianga. Named after Hei, a founding ancestor of the Ngāti Hei people, Hahei is derived from Te-Whanganui-A-Hei (or Great Bay of Hei), the Maori name for nearby Mercury Bay.

During the late 1970s the Thames-Coromandel District Council sought to rezone land on the south headland of Hahei beach (shown here) to enable its subdivision for 20 holiday homes. RAEWYN PEART

Hahei has a lovely white sandy beach stretching over more than a kilometre and sheltered waters for swimming. It is the gateway to the spectacular limestone caves at Cathedral Cove. During the 1960s, the land behind the beach had been subdivided into a small seaside holiday settlement of around 300 houses.

The southern headland of the beach, which ends at Hereheretaura Point, is a long arm of land extending some distance out into the sea. Framing the beach, it contains the remains of two significant Ngāti Hei pā sites. Although most of the headland was in private ownership as part of the Te Pare Block, it had remained undeveloped. But that was soon to change if council and the land owner had anything to do with it.

EDS director Malcolm Bowman recalls,

> we had a section at Hahei and on Easter Saturday, with Gary and Mary [Taylor], we walked over the hill and stumbled on a ratepayers meeting. The chairman of the ratepayers association had invited the developer from Thames to discuss the development of the southern headland called Te Pare.
>
> We were sitting outside and the developer had a flip chart with all the plans on it. These included a beautiful artist's rendition of what the development would look like, showing the headland dotted with houses. The chairman said 'ladies and gentlemen thanks for coming today and resolving the ambiguity regarding the possible development of the headland. Let's give thanks to Mr Smith.'
>
> I then stood up and said, 'excuse me, but some of us think it's not a good idea'. Gary then said, 'I am the executive secretary of EDS and I think this violates the Town and Country Planning Act and we will look into it.' At that point everything went silent and the developer took down his easel and went back to Thames. The meeting disbanded and EDS picked up on the issue.

In November 1979, the Thames-Coromandel District Council resolved to initiate Change No. 3 to its district plan to authorise development of the headland for 20 holiday homes. The land had been zoned 'coastal' in the district scheme, the protective coastal zoning that EDS had earlier fought so hard to secure. The council now wished to remove this protection to rezone part of the headland as the 'Hahei Holiday Residential Zone'.

Laurie Newhook lodged an objection on behalf of the Physical Environment Association of the Coromandel. This focused on the 1973 amendment to the Town and Country Planning Act which required district schemes to protect the coastal environment from 'unnecessary subdivision and development'. When the council dismissed the objection, Newhook filed an appeal. Just before the hearing in front of the Town and Country Planning Appeal Board, Newhook discovered he had a conflict of interest with a previously undisclosed development partner, so Coromandel resident Mary Foreman, who was chair of the Physical Environment Association, presented the case that he had prepared.

The association's case was supported by evidence from a Māori elder that development of the headland would be contrary to the culture and traditional relationship that Māori had with the land. However, this evidence was dismissed by the appeal board on the basis that the land was no longer owned by Māori. (It wasn't until 1987 that EDS managed to overturn this narrow approach to Māori ancestral land in the High Court, when acting for Forest and Bird on a case about mining on the Kaitorete Spit in Canterbury.)

In its decision, issued in June 1982, it was clear that the appeal board was persuaded by the argument presented by the Physical Environment Association. Its decision stated,

> Headlands are important features in the coastal environment. They are important because of the close relationship between hill and sea. They are also important because they enclose the environment when viewed from a distance, and because they are an obvious part of what is perceived ... because of the importance of the headland in the coastal environment, only the most compelling need would justify subdivision and development of any part of the Te Pare block.

No such compelling need was demonstrated in evidence, so the Town and Country Planning Appeal Board turned down the change to the district scheme, and the headland remains undeveloped today. As Malcolm Bowman observed, 'if we hadn't stumbled on the ratepayers meeting at that particular time and spot on the earth, we wouldn't have known about the proposed development and have been able to help stop it'. The case established a precedent on the importance of protecting headlands from development which had application everywhere along the country's extensive coastline.

AOTEA GREAT BARRIER ISLAND, located some 90 kilometres from Auckland, is the largest island in Tīkapa Moana o Hauraki Hauraki Gulf. It is the ancestral home of the Ngāti Rehua–Ngātiwai ki Aotea hapū of Ngātiwai. The landscapes of the island are predominately volcanic, which gives rise to a spectacularly rugged landscape and a large variety of coastlines. The west coast is covered

Judge Laurie Newhook graduated with a Bachelor of Laws (Hons) from the University of Auckland in 1972. In 1976 he became a partner at law firm Towle and Cooper (which later became Brookfields), with a practice in resource management, local government and general litigation. In 2001 Newhook became a judge of the Environment Court and was appointed its head in 2011. He retired as Chief Environment Court Judge in 2020 but still acts as an alternate judge on that Court. In 2022 he was appointed Chief Freshwater Commissioner and was appointed by the Minister for the Environment to be Convenor of Expert Panels under the Covid-19 Recovery (Fast-track Consenting) Act 2020, for the three-year life of the Act.

Newhook was a founding law student director of EDS. He remained on the board until 1973, when he became an advisory counsel to the society. When reflecting on his involvement in the Coromandel Peninsula cases for EDS, Newhook recalls, 'there I was, a newly minted lawyer appearing against none other than David Sheppard for the council and Peter Salmon for the developers on the peninsula. And I won some, which was pretty satisfying. I got to run litigation well beyond my experience and more than my then employer was offering me. It was a very welcome part of my early experience in environmental law. I thoroughly enjoyed it, and it was satisfying to achieve environmental protection in that lovely part of New Zealand.'

In 2021, Newhook was awarded the Resource Management Law Association's supreme award (last awarded to Hon Peter Salmon KC in 2017), the R. J. Bollard Lifetime Commemorative Award for 'outstanding lifetime services and contribution to the advancement of resource management and environmental law in NZ and securing the place of the Environment Court within the NZ judicial system.'

COURTESY OF LAURIE NEWHOOK

Newhook has also been co-opted by the United Nations Environment Programme onto a steering group which is examining the concept and feasibility of a regional environmental dispute-resolution body for the Pacific.

Great Barrier Island has largely escaped the intensive coastal subdivision that affected the Coromandel Peninsula from the 1950s. Shown here is the small settlement of Port Fitzroy, which contrasts with the rugged undeveloped hinterland. RAEWYN PEART

in steep forested ranges that lead down to sheltered bays and islands drenched in pōhutukawa. This contrasts with the exposed and windswept east coast that has long, white-sand beaches, extensive dune systems, tidal creeks and wetlands.

The island had largely avoided the coastal development gold rush of the 1950s and '60s that had hit areas such as the Coromandel Peninsula, due to its geographic isolation and difficult access. But that changed in the early 1970s when air services started improving. Islanders became aware of the marketable value of their rural land and began to look at the potential for subdivision and residential development. Land speculation started to occur.

In mid 1974, the Great Barrier Island County Council advertised its first provisional district scheme for public objection, which had been prepared by the Ministry of Works and Development. On examining the scheme, EDS found it to be deficient in many respects and so the society lodged extensive objections. EDS also wrote to the Minister of Works and

Development, Hugh Watt, suggesting that a fundamental reassessment of the planning objectives for the island needed to take place.

EDS was concerned that 'a repetition might occur on Great Barrier Island of the type of subdivisional development that had already ruined much of the finest coastal scenery on the nearby Coromandel Peninsula'. In the society's view,

> the very rapid pace of development has already produced very undesirable consequences, by initiating a progressive urbanisation of the island, which, if unchecked, will destroy not only the unique wilderness and scenic attributes of the island, but will also result in the fragmentation of titles and reduced likelihood of the continued existence of viable agricultural production.

EDS was concerned about the adequacy of the proposed district scheme in dealing with such pressures, and the lack of planning resources within the council, with the population of the island in the 1971 census numbering just 267. At that time, the council chambers consisted of an old tin shed on the edge of the Kaitoke airfield.

EDS also suggested that an interdepartmental committee be established to consider the future of the island and that its potential as a national maritime park be considered. Dick Bellamy remembers meeting Hugh Watt in the corrugated iron shed that served as the Onehunga Labour Party rooms to press EDS's case. There were pictures of New Zealand's first Labour Prime Minister Michael Joseph Savage on the walls. As Bellamy explains, 'in those days it was easy to see a cabinet minister if you were polite and not a danger to the individual. I explained to him the situation of the council, which was hopelessly under-resourced.'

Hugh Watt grasped the nettle. In 1975 he established a one-man Committee of Inquiry, in the person of J. Granville, to study all aspects of the future development of the island. EDS made extensive submissions to the committee, which were prepared by Dick Bellamy and Mike O'Sullivan. They stressed the desirability of protecting the island against development and argued that the district plan should not be progressed until the island's administration was brought under a more suitable body, such as the Rodney County Council. However, this did not happen for some time.

While the future of Great Barrier Island was being debated, EDS was also seeking to head off a development proposal for a remote undeveloped beach on the island's east coast. Harataonga is a circular semi-enclosed bay sheltered to the east by Dragon Island and Rakitu Arid Island. It has a curved golden-sand beach backed by dunes and ancient pōhutukawa trees. A river flows over the sand at the eastern end of the beach and rocky reefs rich in marine life stretch out into the bay. The land behind the beach had been farmed by the Overton family for many years.

In the early 1970s, the Overtons sold their farm to an American developer who had plans for a series of residential developments on the ridge behind the bay. He engaged landscape architect Brian Halstead, who was at that time in partnership with Harry Turbott, to design the development. Turbott was well-known for his sensitive designs, including visitor centres at Lake Waikaremoana and the Waitangi Treaty Grounds.

When Dick Bellamy first heard about the proposal he was horrified. His father had known the Overton family and Bellamy had spent

Harataonga beach in 2008. During the early 1970s there was a proposal to develop housing on the ridge behind the bay (the grassed area in the centre of the image). CRAIG POTTON

time camping at Harataonga. Bellamy rolled into action and embarked on a vigorous letter-writing campaign targeting various ministers and members of the Nature Conservation Council in order to build up a large paper trail in Te Whanganui-a-Tara Wellington.

Fortunately, Don Miller, who was the local planning boss at the Lands and Survey Department, was receptive to Bellamy's concerns. Miller did not favour that sort of development on Great Barrier Island and liked the idea of public acquisition. Bellamy started floating the idea of the whole island becoming a national park on the English model, where national park status applied to private land. As he recalls, 'the national park proposal appeared on the front page of the *New Zealand Herald* and for some time I did not feel safe enough to buy a ticket on the light aircraft that flew out to the Barrier, so it was an interesting time.'

Eventually, the developer decided it was getting too hard and expensive to proceed with the resort in the face of the opposition that had built up. He sold the land to the Department of Lands and Survey and it is now part of the Aotea Conservation Park. There is a popular DOC camping ground behind the beach.

After the Commission of Inquiry reported back in 1975, the Great Barrier Island County Council sought to revoke its entire district scheme and replace it with a new scheme prepared by a Rodney County Council planning officer. It proposed to do this by way of a variation procedure under the Town and Country Planning Act 1953. This was designed to enable councils to change parts of their proposed district schemes before finalisation, when new matters had arisen, but not to

Once the development proposal for Harataonga had been headed off by EDS and others, the land was bought by government and is now part of the Aotea Conservation Park with a popular camping ground, shown here, behind the beach. RAEWYN PEART

replace the entire document. EDS appealed the variation to the Town and Country Planning Appeal Board. Peter Salmon took the case on behalf of EDS.

EDS submitted that the proposed variation was not a variation at all as it completely replaced the original scheme. In EDS's view, it should go through all the procedures applicable to a new scheme. In addition, the variation failed to distinguish or protect land of high agricultural productivity, provided inadequate standards for subdivision and building, and allowed the council too much discretion to make decisions on development proposals on an ad hoc basis. In particular, the variation did not explicitly provide for matters of national importance, including the preservation of the natural character of the coastal environment.

The hearing itself lasted little more than 15 minutes. Peter Salmon was invited to present EDS's case first. He drew on the definition of 'variation' in the *Concise Oxford Dictionary* and the hearing was adjourned shortly afterwards. It was an open-and-shut case.

The appeal board agreed that the county council could not substitute an entire district scheme by way of variation. In addition, it found that the district scheme the council was seeking to put in place was so deficient that it needed to be thrown out. Its decision stated that the variation was 'so deficient in the content or detail required to be included in a district scheme that it is invalid as a matter of law; or if not so invalid that it should be rejected as a matter of discretion.' The only remedy was to start afresh.

It had taken almost a year's worth of revenue for the council to prepare the 'variation' and the plan had now been thrown out. At that point the Ministry of Works and Development decided that the situation was unsustainable. It abolished the Great Barrier Island County Council and put the island under the jurisdiction of the Rodney County Council.

On reflecting on the case, Peter Salmon recalled, 'EDS thought the council had made such a mess of things that it filed an application to set the whole scheme aside. So the council had to start all over again. That's the only case I know of where a total plan has been set aside because of its deficiencies.' This was not to be the first time that EDS was to hold councils to account in notifying grossly deficient planning documents.

EDS kept pushing for the rest of the northern end of the island to be acquired by the government. As Dick Bellamy explains,

> it was evident to me and everyone else that the southern end of Barrier was already seriously compromised, because subdivision had been approved at Medlands Beach, and Tryphena and Ōkupu were already heavily carved up. But the northern two-thirds of the island could be saved from development if we drew a line from Harataonga Bay across to Whangaparapara.

Much of the northern part of the island was owned by Max Burrill, a wealthy South Auckland dairy farmer. After a serious accident on his Auckland farm in 1983, Burrill decided he had had enough and he offered to sell his Great Barrier Island farm to the Department of Lands and Survey. The Crown approached EDS to help raise funds for the purchase price of $2 million. The land was transferred in 1984. It included a gift from the Burrill family of 3600 hectares in addition to the farm itself. The land now forms an important part of the Aotea Conservation Park.

It was not until 1983 that a new proposed district scheme for Great Barrier Island finally emerged. It was much improved. However, EDS was still actively involved in strengthening protection for the island's natural features and, in particular, its native vegetation. EDS argued that due to the environmental and aesthetic significance of the island, special controls should be placed on all developments to ensure that the unique qualities of the island were not compromised. EDS continues monitoring planning provisions for Great Barrier Island to this day.

WHEN GARY TAYLOR first heard about the proposal to develop the Karikari Peninsula in 1978 he thought 'it was outrageous what they were wanting to do, to put a Gold Coast-type high-rise development on the remote Karikari Peninsula. So EDS got involved.' In fact, it was this audacious proposal that helped convince Taylor to give up being a journalist and work for EDS in the first place. It would also attract some of the finest legal minds in the country to contest its merits, and the applicable law, all the way through the courts.

The Karikari Peninsula is situated on the east coast of Te Tai Tokerau Northland, about 25 kilometres from Kaitaia. It consists of a

The Karikari Peninsula, a remote and stunningly beautiful place on the east coast of Northland and is of great significance to Ngāti Kahu, was the site of a proposed holiday resort accommodating up to 10,000 people.
CRAIG POTTON

long narrow strip of land that separates the expansive Rangaunu Bay to the west from the wide, sweeping Doubtless Bay to the east. The peninsula itself features a diverse and wild coastline. There are open coastal beaches at Puwheke and Karikari, rocky headlands at Cape Karikari and Knuckle Point, small sheltered coves at Whatuwhiwhi and Maitai Bay, and the long stretch of Tokerau Beach which forms the western side of Doubtless Bay.

The peninsula is of great significance to Ngāti Kahu, who have mana whenua over the area. Tradition tells of it being the landing place for the canoe that brought their ancestors to Aotearoa, and it was where the iwi first settled.

During the late 1970s, when the resort proposal first came to light, only around 300 people lived on the peninsula, mainly in small Māori communities. It was an isolated rural area dependent on a long unsealed road for access. It was here that Glendene Developments Limited (subsequently renamed Doubtless Bay Development Company) proposed to build a large tourist resort catering for thousands of guests.

The development company had acquired an 800-hectare block of farmland behind Karikari beach. This was to house a resort that would provide a 'full range of accommodation for 6000 to 8000 guests' and offer 'a wide range of entertainment and recreational facilities'. In the Mangonui County Council's district scheme, which had only become operative in 1976, most of the block was within the coastal conservation zone. This was similar to the coastal zone that EDS helped establish on the

Coromandel Peninsula. It was designed to keep coastal areas in their natural state.

Clearly, if the development was to go ahead, the district scheme would need to be changed. The company wanted some certainty that this would happen before investing money in investigating the site and coming up with a detailed proposal.

The company persuaded the council to notify a change to its district scheme to indicate 'in principle' support for a proposed tourist development on the site, and to commit to making further changes to the district scheme to facilitate it after the developer's investigations had been completed. It was this change (Change No. 7) that EDS and others objected to and sought to have abandoned. But the council, being strongly supportive of the development, largely dismissed the objections.

The development also had government support. As Gary Taylor wrote in the March 1979 edition of *EDS News*,

> It is ironic that Lands and Survey, the Government Department that is supposed to ensure the wise management of our coastline, seems to favour the development. Reliable sources have informed EDS that the Department was party to the discussion that preceded the scheme change application and positively encouraged it. In addition, Air New Zealand and Mt Cook Airlines – two public corporations – have an involvement in the proposal. No doubt they are eyeing the potential extra revenue that an increase of flights into the North could bring.

EDS, and others, lodged appeals with the Planning Tribunal (the renamed Town and Country Appeal Board). Auckland barrister Trevor Gould represented EDS along with the Nature Conservation Council in the five-day hearing at Kaitaia. EDS director and planner Kevin Williams gave planning evidence. Peter Salmon acted for the council and David Sheppard acted for the development company. Judge Turner was presiding. The tribunal largely found in favour of EDS and others in removing any specific reference to council support for a resort on the Karikari Peninsula, although not going so far as to remove the scheme change entirely.

In its decision, released on 1 October 1979, the tribunal made it clear that the council should not be cosying up to a developer, something that small councils were prone to at the time. The tribunal stated 'a council should not commit itself in its district scheme to bringing down a change to that scheme *for the benefit of a private landowner*, even if there are conditions precedent to the making of the change'.

This turned out to be only round one of what would become an extremely long and protracted battle over the future of the Karikari Peninsula. The development company went away and undertook studies of its land. In September 1983, when the council notified a review of its district scheme, no provision was made for the Karikari resort. But shortly thereafter the company released details of a much more developed proposal. It now comprised three stages.

Stage 1 included 200 hotel rooms, four motels, a serviced camping ground, an international-sized golf course with 200 accommodation units, commercial buildings, an equestrian centre and recreational facilities, such as squash and tennis courts.

Stage 2 included a further 200 tourist units, artificial lakes and up to 300 lakefront units, a tourist village with up to 400 tourist units, up

To support EDS and Forest and Bird in fighting the Karikari Peninsula development, artist Don Binney contributed a silk-screen original artwork depicting the northern shores of the peninsula. Numerous posters of the image were sold to raise funds for the campaign. COURTESY OF MARY BINNEY

to six more motels and 200 more hotel rooms, and a further camping ground.

And stage 3 included a third camping ground, 300 more lakefront units, another tourist village with up to 400 units and up to 200 more hotel rooms. By the end of stage 3, the development would provide for more than 10,000 guests.

As there was an extensive wetland behind Karikari beach, the main development was to be located 1 to 2 kilometres inland, on higher ground. There were plans to transport tourists to the beach by vehicles and to develop facilities for them there.

There was nothing else like it in the country at the time, and nor is there now, nearly 40 years later. When the company came back with the developed plans, council was very supportive. Such a development could help generate much-needed jobs and economic development in an impoverished region. But it also imperilled an unspoiled wilderness area.

EDS, with the support of Forest and Bird, set up a 'fighting fund' to oppose the development. Notable New Zealand artist Don Binney contributed a magnificent silk-screen original artwork showing a view of the northern shores of the Karikari Peninsula. Posters of the image were sold to raise funds for the campaign.

In May 1984, the council notified a variation to its district scheme (Variation 1) to rezone land for the development. EDS, the Tai Tokerau District Māori Council and others objected. The objections were heard by the Mangonui County Council in September 1984 with the hearing given detailed coverage in the local media. This made it clear that emotions were running high.

On the first day of the hearing, written submissions were presented on behalf of Sir Graham Latimer for the Tai Tokerau District Māori Council. He expressed deep concern about the impact of the proposal on the small local Māori community, stating 'at the present moment all I can foresee is that if you put $27 million into Whatuwhiwhi [a small settlement on the Peninsula] you will create a flood that will carry my people out to sea'.

Gary Taylor presented extensive submissions on behalf of EDS, arguing that there was no need for the resort, and even if there was, the Karikari Peninsula was not the best place for it. Such a large resort should be located in an area already compromised by development, he contended.

Taylor also spoke of the need to protect the peninsula in its present wilderness state, and of the great significance of the area in terms of wildlife and geological interest, including supporting a small breeding population of rare New Zealand dotterels (tūturiwhatu).

As well as suggesting that the project should be deferred until a full environmental impact report had been prepared, Taylor called for a comprehensive social impact assessment, along with a study of the development's impact on the Māori community.

EDS's opposition to the proposal was not well received by its local promotors. In his submission, Kaitaia Mayor Desmond Bell opined,

> local people deeply resented outsiders coming in and trying to upset a scheme which would bring employment and money to the north ... the Environmental Defence Society had stated that its membership consisted mainly of university academics and professionals who enjoyed the wilderness aspects of Karikari on their holiday visits. I could not help being amazed at the arrogance of this self-styled elitist group with their substantial and secure incomes, who apparently have no concern for the people of the Far North.

Such sentiments were echoed by the developer. The Minister of Works, Fraser Colman, who had initially called for the withdrawal of the proposal, now indicated that his concerns could be addressed if the variation was modified to pull buildings well back from the shore, follow land contours and create inland water areas.

In response, the council adjourned the hearing and instructed a planning consultant to report on the matters raised by the minister. This led to a second variation (Variation 4) being notified by council which amended the provisions of Variation 1. EDS and others objected to the new variation, but again their objections were dismissed by the council, which approved the change to the plan, thereby paving the way for the development to go ahead.

As reported in the February 1985 edition of *EDS News* under the headline 'Karikari: Think big', 'The Karikari development will be the biggest single coastal development in New Zealand history! Ultimate visitor capacity will be 10,000 – more than the total of people living in Mangonui County.'

At that point EDS appealed to the Planning Tribunal with the Tai Tokerau District Māori Council in support. The hearing was held in Kaitaia during September 1985. This time Russell McVeagh lawyer Gerard Curry was acting for EDS, assisted by Deborah Clapshaw. Peter Salmon, now a Queen's Counsel, was acting for the council and Alan Galbraith (who later became a Queen's Counsel) represented the development company. Sian Elias, who would later become the country's Chief Justice, appeared for the Tai Tokerau District Māori Council.

EDS called several witnesses to support its case. A planner/architect who had considerable experience in planning for tourism said the enterprise would likely fail due to its size and location and absence of supporting tourism ventures. He was concerned that 'the outcome of an unsuccessful development would be an irretrievably altered natural environment and something requiring considerable support or subsidy or being left to deteriorate into an unattractive coastal settlement'. In effect, the development could become a decaying 'white elephant'. In his view the land should be retained in its present rural/natural state and ideally be acquired by the Crown.

Another witness for EDS, who had experience with tourist resort development overseas, said that the main attraction at Karikari would be the beach and marine environment and therefore tourists would demand to be in easy walking distance of the beach. Concentrating the development 1 to 2 kilometres away from the coast was, in his view, 'absurd in resort terms'. An economist and consultant town planner gave evidence on the likely social and economic impacts of the proposal. And finally, an expert ecologist talked about the high values of the sensitive dune vegetation which was important for the summer breeding of the New Zealand dotterel.

In support of EDS, a Ngāti Kahu spokes-person gave evidence that although his people wished to see progress, they did not support a large-scale tourist resort. His people would like to be associated with a smaller, locally related development which would fit into their community and with which 'they could grow into a warm, welcoming and distinctively different tourist centre.' An archaeologist also gave evidence on the impact of the development on Māori archaeological sites.

In the end, the Planning Tribunal was persuaded that there was a place for such a destination resort in New Zealand and that it should be in Northland. It also found that the resort might substantially reduce local unemployment and under-development with few apparent alternatives. It said little about Māori cultural concerns.

The tribunal authorised the scheme variation, but only to provide for Stage 1 of the development. This meant that visitor capacity would be a more modest 2280 rather than close to 10,000. It also excluded any development on the beach itself and adjacent wetland meaning that the wetland would be protected from the reclamation and excavation originally proposed by the developer.

The February 1986 edition of *EDS News* claimed that the tribunal's decision

> vindicates EDS's opposition to the development and calls into question the role of both the Ministry of Works and the Nature Conservation Council. MWD strongly supported the proposal, ignoring both the environmental impacts of such a huge proposal and the concerns of the local Maori community. Nature Conservation Council refused to get involved in a public interest role, indicating very clearly the conservatism of that organisation and an apparent inability to act independently of MWD. In spite having been reduced in size, the Karikari development will still be the biggest resort in New Zealand and perhaps the South Pacific.

Although the developer did not get all that it had sought, EDS had lost its bid to protect the remote area from development. But the society was not about to give up. It lodged an appeal to the High Court, as did the Tai Tokerau District Māori Council. The hearing of the appeals in front of Justice Chilwell commenced in late March 1987.

It was now over a decade since the possibility of a resort at Karikari had first been mooted. It was the longest-running case that EDS had been involved in. Gerard Curry and Paul Majurey acted for EDS, Sian Elias for the Tai Tokerau District Māori Council and Peter Salmon for the council and now also the development company. The Minister of Works and Development was also represented at the hearing in support of the development.

Appeals from Planning Tribunal decisions were only possible on matters of law, and could not challenge the merits of the matter, so the appeals were focused on two key legal points. EDS's lawyers argued that the Planning Tribunal had not given sufficient weight to the 'preservation of the natural character of the coastal environment ... and protection ... from unnecessary subdivision and development'. This was specified in the Town and Country Planning Act 1977 as a matter of 'national importance' which district schemes must recognise and provide for. EDS argued that where a development would negatively impact on the natural character of the coast, the need for the development must be established before it can go ahead.

No evidence on need had been provided to the tribunal, nor any market or feasibility study. There was also no indication as to whether any party had the financial where-with-all and interest to undertake the development. But the tribunal had not deemed such evidence to be necessary. Instead, it had largely focused on the suitability of the site for such a development and was swayed by its possible benefits for the local economy.

The tribunal had effectively undertaken a 'balancing exercise' between the national imperative to protect undeveloped coastline and the local need for economic development. In doing so, it had concluded that the local matters should be given more weight and should prevail. Such a balancing approach had been the norm since the Town and Country Planning Act had come into force a decade earlier. But EDS thought it was wrong in law and was now challenging the last 10 years of legal jurisprudence. This was a step too far for Justice Chilwell who upheld the tribunal's balancing approach.

The lawyers for the Tai Tokerau District Māori Council focused on a separate matter of national importance which referred to 'the relationship of the Māori people and their culture and traditions with their ancestral land'. Their lawyer, Sian Elias, explained that only about 150 people of Ngāti Kahu descent lived on the Karikari Peninsula and, of these, only about 40 were of employable age. The sheer scale of the development would overshadow their presence in the area and 'they could not relate to or feel at home with a development of this kind and magnitude'.

The tribunal had not made a finding as to whether the development site was ancestral land and in fact made no reference to the relevant section of the Act. Despite this, Justice Chilwell found that the tribunal had considered the impact of the development on local Māori and that was sufficient.

Both appeals were dismissed, effectively giving the development the go ahead. EDS and the Tai Tokerau District Māori Council sought leave to appeal to the Court of Appeal. After the High Court hearing, the Ministry of Works and Development was disestablished and the new Minister of Conservation (heading the new

Maitai Bay, Karikari Peninsula, 2008. EDS and the Tai Tokerau District Māori Council opposed the resort development on Karikari Peninsula for more than 12 years and finally prevailed when in 1989 the developer abandoned the proposal. CRAIG POTTON

Department of Conservation) took over the government's interest in the case. This resulted in a change of approach, with the government now opposing the development and supporting the appeals.

The same legal team argued the case in May 1988 in front of five judges in the Court of Appeal. It was the first time that the application of the matters of national importance under the Town and Country Planning Act had been considered by that court.

It was clearly a difficult legal matter because the Court of Appeal judges disagreed over which approach should be adopted. In a February 1989 decision, a narrow majority of three judges against two found in favour of the EDS and Tai Tokerau District Māori Council appeals. They did not definitively turn down the proposal but remitted it back to the Planning Tribunal for a full rehearing. It was now 1989, some 12 years after the first development proposals for the peninsula had been revealed, and at this point the developer threw in the towel. He passed away in the early 1990s and the land was put on the market. It was sold to an American merchant banker. And, as it turned out, this was only the first salvo in attempts to develop the peninsula.

The Court of Appeal decision overturned more than a decade of case law which had 'balanced' national and local considerations. It held that matters of national importance must be given greater weight and the developer needed to show that the development was necessary and outweighed other national interests.

The decision also highlighted the importance of protecting the relationship of Māori with ancestral land. Allowing Stage 1 of the development to take place, accommodating 2280 people, when the present population of the peninsula is only a few hundred 'could open the floodgates and see the remaining Ngāti Kahu people with their culture and traditions swept away'.

Over 30 years later, Chief Environment Court Judge David Kirkpatrick observed, 'the Mangonui case remains a landmark decision to this day. It is to do with old legislation but it's still something to guide you. If I were asked to identify EDS's most notable case, I would think of Mangonui County first.'

BEFORE THE TOWN AND COUNTRY Planning Act went out of existence, the Karikari case proved to be decisive in another battle, this time fought back on the Coromandel Peninsula at Opoutere. Opoutere is located on the eastern coast of the peninsula, a short distance north of Onemana and on the edge of the Wharekawa Harbour.

On the north side of the harbour is a sandspit which had been declared a wildlife reserve to protect the breeding sites of the threatened New Zealand dotterel and variable oystercatcher. The beach stretches northwards from the tip of the sandspit to form a long, wild, white-sand surf beach. Unlike other popular beaches on the Coromandel Peninsula, only a few baches had been built next to the harbour, and none behind the beach, which was backed by reserve land. It was one of only two major beaches on the peninsula without urban development behind the beach (the other being Waikawau Bay described earlier).

New Zealand Waterways Association Limited (later called Opoutere Park Beach Resort Limited) owned a pocket of land between the Opoutere village and the sandspit and was keen to develop it for a camping ground. As well as providing camping sites, the company wanted to build up to 24 cabins and associated service buildings. The land was within the coastal protection zone. There were many objections to the proposal, including from the Opoutere Residents and Ratepayers Association. Through EDS, Ian Cowper took the case on for the association.

In the first instance, the Thames-Coromandel District Council granted consent, but for a reduced development of up to 12 cabins. The Opoutere Residents and Ratepayers Association appealed, seeking to reduce the permissible development. The developer also appealed, wanting to increase the number of cabins back up to 24.

In an interim decision released in October 1985, the Planning Tribunal was not sympathetic to the residents' concerns and it granted consent to the developer. As in the Karikari case, the tribunal did not give the national imperative to protect the natural character of the coastal environment from 'unnecessary' development any primacy but considered it as merely part of a mix of factors to be taken into account and ranking equally among themselves. In particular, although the tribunal did not find that the camping ground

Wharekawa Harbour, 2023. Opoutere is one of the few remaining places on the Coromandel Peninsula where there is no urban development behind the beach. EDS supported the Opoutere Residents and Ratepayers Association to oppose the development of 24 cabins as part of a camping ground at the head of the harbour.
RAEWYN PEART

development was 'reasonably necessary', it still granted it consent.

The residents then appealed to the High Court but were again unsuccessful. So Ian Cowper decided to take it all the way to the Court of Appeal. By the time the case was heard in that court, the Karikari Peninsula decision was pending. The court's decision on Opoutere was issued shortly after the Karikari one. The same reasoning was applied, the Planning Tribunal's decision was set aside, and the matter sent back to the tribunal for rehearing. In the end, the development was approved but at a significantly smaller scale, with only six cabins set back from the end of the spit permitted.

Sadly, the import of the Karikari decision was to be short-lived as, just two years later, the Town and Country Planning Act was swept away by a reforming government and replaced by the Resource Management Act 1991. And although much was carried through into the new legislation, including matters of national importance, the reference to 'unnecessary' subdivision and development in the coastal environment, which had saved the Karikari Peninsula, was expunged from the statute book.

8. End of an era

BY THE MID 1980s, EDS was starting to run out of steam. The organisation had achieved an enormous amount over a 17-year period. But this had been reliant on the considerable energy and volunteer effort of a small group of lawyers and scientists who also had busy professional careers. As former EDS director Mike O'Sullivan observed, 'it was damn hard work being an activist so typically people ran out of steam after a year or two'.

IN 1984 GARY TAYLOR resigned as the EDS executive officer when he decided to run on a Labour ticket for the Hauraki seat in the general election. This was in order to raise the profile of opposition to mining on the Coromandel Peninsula. In recognising Taylor's contribution to the society, EDS founder David Williams wrote:

> It will be extremely hard for EDS to find someone who possesses the qualities that you brought to the role: A deep commitment to the cause of conservation, an ability to articulate the environmental viewpoint most persuasively and an operational pragmatism which enabled EDS to take effective action. We are all very much in your debt and I find it very hard to think of EDS without you.

As it turned out, Taylor's electoral bid failed, which was not a surprise as Hauraki had long been a safe National seat. However, his efforts did serve to raise the profile of the mining issue within the electorate.

Taylor then continued to work for the society on a part-time consultancy basis, but his considerable energies were now dispersed. At the same time as he was working for EDS, Taylor was a councillor and then deputy mayor of the Waitematā City Council and also, for a time, chairman of the Auckland Area Health Board and the works committee of the Auckland Regional Authority. Others were employed by EDS from time to time, including Kevin Williams who worked as a research officer for several months and Marjorie van Roon, who briefly took over the role of executive secretary.

EDS had been a largely male-dominated organisation but some women were appointed to the board. The most notable were Jeanette Fitzsimons, who later became a long-standing Green MP and co-leader of the Green Party, and Dame Sian Elias KC, who became Chief Justice.

Funding the organisation had been a constant struggle. The agreement with Forest and Bird to provide financial support to EDS in exchange for legal assistance worked well for some years, but it came to an end in early 1985 when Forest and Bird came under financial strain. The loss of this funding was a severe financial blow for EDS, as it had comprised around two-thirds of the society's regular income. EDS lost its physical presence in mid-1985 when, unable to afford the rent, it closed down its town hall office.

The dire situation was described by then EDS Chairman Dick Bellamy in the July 1986 issue of *EDS News*.

> Financial survival of the Society over the past twelve months has required the closure of the central office, disconnection of the telephone and decentralisation of our work-load by distributing tasks to individual directors and volunteers. But we are unable to escape the costs of xeroxing, toll calls, transport and disbursements to those witnesses and advocates involved in travelling to attend hearings.

One project that was providing income for EDS at that time was a contract to undertake a pilot study into environmental legal aid.

Previous pages: Looking out over Okahu Bay and Tāmaki Makaurau Auckland CBD. RAEWYN PEART

Funded by the Ministry for the Environment, the scheme involved EDS providing assistance to groups and individuals seeking legal representation and advice in the northern part of the North Island. EDS also identified the need for an environmental defenders office that would distribute legal aid funding, offer advice and provide advocacy to individuals and groups. The society had found that being a public environmental defender with no secure income source was simply too hard to maintain over the long term. In 2000, the government established an environmental legal aid fund which is still operating. EDS continues to advocate for the establishment of an environmental defenders office.

One solution to EDS's financial woes was to combine efforts with those of another organisation, in this case the Native Forests Action Council. The two organisations appeared to have complementary skill sets. Having embarked on the largely successful campaign to stop the logging of native forests, the Native Forests Action Council had a much wider group of supporters and larger funding base than EDS. On its side, EDS brought formidable legal skills. A merger took place in May 1988 to form a new entity called the Maruia Society. At that point EDS, as a separate entity, went into recess.

Part of the rationale behind the merger was that the forthcoming reforms, which were to radically change environmental laws and institutions and provide for greater public accountability and environmental protections, would reduce the need for a public environmental defender such as EDS. However, that was to prove wrong.

New Zealand had become a very different country since EDS was first formed in 1971. At that time, environmental law was in its infancy and there was virtually no public involvement in government decision-making affecting the natural environment. Laws were largely designed to promote industry rather than to curb environmental impacts. Government considered itself to be largely outside whatever rudimentary environmental controls were in place. And when it came to environmental matters, local councils were largely moribund, and easily captured by development interests.

The efforts of EDS helped to change all that.

In the early years, EDS was able to make considerable impact due to the novelty of its approach. As Mike O'Sullivan observed,

> EDS was one of the first really effective lobby groups. Some of the old environmental groups had been around for a while, like Forest and Bird, but they tended to be quite passive then. There was a golden period of four to five years when companies and government didn't know what had hit them, and that made it possible for EDS to be really effective. But by the late '70s, the establishment had started to become more sophisticated in dealing with EDS and opposing what we had to say. So life for EDS became harder.

Dick Bellamy recalls,

> in the late 1960s and 1970s New Zealand was just emerging from a period where almost all development that was proposed was assumed, automatically, to be a good thing. Government departments such as the New Zealand Electricity Department and the Ministry of Works and Development were major drivers of national and regional development and the government controlled almost

> everything else as well. So opposition to a project was really viewed as opposition to the government's intentions.
>
> Then, it was considered very radical to challenge the government and its major ministries because they felt they were, by definition, working in everybody's best interest. They found it difficult to understand why their favourite projects and poorly founded decisions should be challenged, particularly if it was in a professional/legal way.

Part of EDS's effectiveness was due to its careful selection of cases. As Tony Randerson recalls,

> if we thought an issue needed to be run we ran it, usually to pretty good effect. The cases were chosen carefully and there was a rule that no case would be filed unless a QC had run their eye over it and said 'yes this is an arguable case'. That was very good protection. From the lawyer's point of view the litigation was novel, was out of the normal run of cases. It was interesting and if you were keen on preserving the quality of the environment it was a worthwhile thing to be doing. There was no shortage of lawyers willing to do it for that reason. But it meant a lot of work out of hours, during nights and the weekends.

The previous chapters highlight some of the more notable cases and issues that EDS became involved in. But there were many more.

EDS helped clean up gross pollution around the country. As well as closing down the Auckland Hospital chimney and holding the Huntly Borough Council to account for its raw sewage discharge into the Waikato River, EDS targeted other examples of poor environmental practice. It lodged private prosecutions against the Auckland City Council for pollution of the Waitematā Harbour from its rubbish tip at Meola Reef and against the Takapuna City Council for discharges from the Barry's Point Road dump. It also stopped the discharge of polluting clay washings into the Whangārei Harbour by Wilsons NZ Portland Cement Limited.

Alongside the Waitākere Ranges Protection Society, EDS helped head off a planned rubbish tip in the Waitākere River Valley, which at the time was one of only two cases where water rights had been refused by the regional water

Whangateau estuary, 2015. One of the numerous pollution cases that EDS was involved in was stopping a proposed sewage treatment site on the shores of the Whangateau estuary. RAEWYN PEART

board. The society also helped stop a proposed landfill at Ōkura and a sewage treatment site on the shores of the Whangateau estuary.

EDS's work, along with the acclimatisation societies, helped embed into law the proposition that natural freshwater systems needed protection. It promoted the passage of legislation to protect wild and scenic rivers, was instrumental in achieving protection of the Mōtū River, and helped set legal precedents to assist with the protection of further rivers. EDS also supported scientists to achieve protection of the Waimangu geothermal field.

EDS made a major contribution to coastal protection during this era. The society helped to strengthen district schemes in key areas under development pressure including the Te Tara-o-te-Ika-a-Māui Coromandel Peninsula, Pēwhairangi Bay of Islands, Great Barrier Island, Waiheke Island, Rodney County and Marlborough Sounds. EDS stymied development at Waikawau Bay, Hahei, Karikari Peninsula, Opoutere and Te Karo and facilitated the extension of the Kumutoto Scenic Reserve in the Marlborough Sounds.

Okahu Bay was once the location of numerous boat moorings, as shown here in 1975.
BASIL WILLIAMS, AUCKLAND LIBRARIES

The society played a key role in heading off mining on the Coromandel Peninsula, thereby saving precious coastal areas such as Otama from despoliation, and was instrumental in bringing mining law into the twentieth century.

EDS lodged the first application for a marine reserve at the Poor Knights Islands in 1972 after being approached by the New Zealand Underwater Association. The association had become alarmed after the Minister of Marine granted an oil prospecting licence around the islands to an Australian company. The EDS application was superceded by one lodged by the Hauraki Gulf Maritime Park Board, which eventually succeeded in creating the reserve in 1981.

EDS was also concerned about government- and private-sector decision-making creating unnecessary risk for the community. It fought hard to reduce the use of the toxin 2,4,5-T and to knock the prospect of nuclear power in New Zealand on the head. It ran a campaign to remove lead from petrol and to incorporate robust safety measures into LPG facilities. It was active in heading off the proposed PVC plant at Marsden Point, which would have used the toxic and highly flammable vinyl chloride gas, and also helped ensure houses built at Matarangi Beach were located safely away from coastal erosion harm.

EDS often worked alongside local Māori: including with Ngāti Mahuta to tighten up conditions around the Huntly Power Station water rights; with the Whātāpaka Marae to head off the Auckland Thermal No. 1 power station on the shores of the Manukau Harbour; and with Ngāti Kahu to oppose resort development on the Karikari Peninsula. EDS also worked with Ngāti Whātua Orakei when the Okahu Bay marina was consented to obtain a commitment to remove the boat moorings from Okahu Bay once the marina was built. In addition, EDS

EDS supported Ngāti Whātua Orakei to get initial agreement to remove the moorings from Okahu Bay once the new marina was built. They were eventually all removed in 2019 enabling the bay to revert to a more natural state. RAEWYN PEART

provided assistance to the Maukarau Marae to successfully oppose the establishment of a Seventh Day Adventist Church complex near by on the shores of the Manukau Harbour.

Much of EDS's work was in support of local communities. Most notably, EDS provided apricot orchardists in Cromwell Gorge and Waihi residents with a voice when their homes and livelihoods were threatened. The society also helped residents successfully oppose a marina proposal at Herald Island in the Waitematā Harbour.

Through constantly challenging legal restrictions on standing, EDS helped widen public participation in environmental decision-making, particularly through the rights to represent the public interest which were inserted into the Town and Country Planning Act 1977. Most importantly, EDS did not shy away from holding the government to account. It did so with the Huntly Power Station case, the Aramoana smelter and the Clyde Dam. EDS also opposed the proposal to construct an ammonia-urea plant in Waimate on the grounds that it was not a good use of natural gas. It brought government to heel in the Nelson Post Office case, where the Ministry of Works and Development illegally constructed a building in excess of the height restrictions in the district scheme.

—ooo—

EDS PLAYED A LARGE role in strengthening environmental law in New Zealand. Environmental law, as such, did not really exist in New Zealand in the early 1970s. The first textbook on the subject was only published in 1980, and this was written by EDS founder David Williams. A seminar on environmental law was first included in the Auckland Law School programme in 1972 and only became a full law degree course in 1974. It was initially taught by EDS founder David Williams and then EDS director Stephen Mills.

EDS organised many seminars and conferences to highlight problems with the law and canvass possible reforms. It co-organised four energy conferences between 1974 and 1978, the first environmental law conference in the country in 1975, a seminar on planning reform in 1980, two seminars on the management of geothermal fields over 1981 and 1982, and a seminar on planning for major resource utilisation in 1983. The society also published a citizen's guide to the Mining Act and to environmental law more generally (*Citizen's Guide to the Law*).

EDS positively influenced much new environmental legislation, including the Clean Air Act 1972, Marine Pollution Act 1974, Town and Country Planning Act 1977, National Development Act 1979, Official Information Act 1982 and Environment Act 1986. It was instrumental in prompting the 1981 amendments to the Mining Act 1971 to strengthen management of the mining industry and to the Water and Soil Conservation Act 1967 to provide for wild and scenic rivers.

EDS also contributed substantially to discussions about the future of the environmental law system leading up to the establishment of the Ministry for the Environment in 1986 and the Department of Conservation in 1987.

As Gary Taylor recalls,

> I think people's recollection of the Muldoon years was that it was the time of unenlightened environmental policy, best exemplified by the National Development Act. But in fact there was a lot of positive stuff happening because there were some relatively enlightened people in the National Party cabinet including Environment Minister Venn Young and his successor Dr Ian Shearer, and Works and Development Minister Derek Quigley. Even Energy Minister Bill Birch had a green tinge.
>
> In 1981 we saw the antiquated mining legislation changed, the wild and scenic rivers amendment pass and the official information legislation opening up government decision-making. It was hard being green during the 1970s, but during the 1980s there was a change in sentiment towards environmental matters in New Zealand. There was still a long way to go, of course, and it was nothing compared to what happened when Labour won a landslide victory in 1984.

There were many others who also worked to improve the environment during this era. For example, the Department of Lands and Survey promoted a network of coastal reserves, the Wildlife Service brought threatened species back from the brink of extinction, the acclimatisation societies sought many more water conservation orders, and the National Parks and Reserves Authority kicked off the Protected Natural Areas Programme,

Meola Reef and adjacent creek. EDS lodged a private prosecution against Auckland City Council seeking to stop harbour pollution from the large council-operated rubbish tip at Meola Reef. RAEWYN PEART

which mapped and described representative ecosystems on private land, providing critical information which is still used today.

As with these efforts, many of EDS's achievements have endured. When the Resource Management Act came into force in 1991, the Crown was explicitly bound by council plans, mining was put under the same level of environmental scrutiny as other activities, there were broad provisions for public participation, provisions to protect the natural character of the coast were retained (albeit with the word 'unnecessary' removed) and the wild and scenic river provisions were carried through intact.

Even when EDS failed in the courts, its efforts often served to highlight issues that resulted in action and sometimes changes in the law for the benefit of the environment.

In holding decision-makers to account, EDS earned well-deserved respect. The society proved itself to be a fearless proponent for environmental justice. As Environment Minister Dr Ian Shearer wrote in August 1981, 'In my view the Society is a responsible, well-respected and effective environmental advocate in this country.'

— PART II —

A new era: 1999 to present

9. EDS reborn

MUCH HAD HAPPENED in the field of environmental law and administration by the mid-1990s. Most significantly, the Resource Management Act (RMA) had passed into law in 1991, radically reforming planning and environmental laws. It brought management of land, air, freshwater and the coastal marine area into an integrated statute under the rubric of 'sustainable management'.

ALONGSIDE ENVIRONMENTAL reforms, there had been a radical restructuring of institutions. The Ministry of Works and Development, the country's national developer and dam builder, had been disestablished in 1988. The Ministry for the Environment, founded in 1986, was given the part of the Ministry of Works and Development's former role that involved oversight of environmental and planning matters.

A year later, the newly established Department of Conservation was charged with managing a third of the country now comprising the conservation estate. It was also tasked with oversight of coastal management and advocating on behalf of natural and historic heritage on private land.

Local government had been restructured to form a national network of regional and district/city councils, with local government entities reduced from more than 625 to just 94.

The new legislation had been developed during a time when neo-liberal thinking was ascendant in New Zealand. The Fourth Labour Government, which was in power from 1984 to 1990, had embarked on a programme of economic deregulation. Central planning was ditched in favour of market-led development. The idea behind the RMA (which was initially developed by the Labour Government but brought into law under the subsequent National Government), as stated by former Environment Minister Simon Upton on the third reading of the bill, was that it would establish biophysical bottom lines that must not be compromised and so long as these were met 'what people got up to was their affair'. However, such bottom lines were slow to emerge and market-led forces drove development.

By the time EDS was re-established in 1999, the RMA had been in force for eight years and much case law had already been established. In particular, councils and the courts had reverted to the old 'balancing approach' to environmental decision-making that had been common practice under the Town and Country Planning Act before EDS's Karikari Court of Appeal decision in 1989.

Without an entity like EDS going to court on behalf of the environment, the bulk of the litigation on the RMA had been initiated by developers, who argued for a more permissive approach to environmental management. And there were few counter-balancing forces. The Department of Conservation had been grossly underfunded from the outset. It soon suffered further budget cuts and pulled back from its advocacy functions.

The Ministry for the Environment, which was charged with developing national policy to guide council decision-making under the RMA, was missing in action. Councils were left largely to themselves when it came to implementing the new statute, and they were far from independent arbitrators. Many regional councils, which were charged with freshwater and catchment management, were politically dominated by the very rural sector they were tasked with regulating. And district councils frequently caved in to pressure from developers to weaken rules in their district plans. There was no one keeping them honest.

As EDS director Graeme Lawrence recalls, 'what happened when the RMA came in was that the early cases were taken by people who

Previous pages: Waikawau Bay estuary, Te Tara-o-te-Ika-a-Māui Coromandel Peninsula. RAEWYN PEART

Opposite: It was the threatened development of Pakiri on the east coast about 100 kilometres north of Tāmaki Makaurau Auckland that motivated Gary Taylor to resurrect EDS. It was the longest undeveloped beach in the region and had very high natural character values. CRAIG POTTON

had the money to go to court. There was no advocate for the new legislation. There was a huge gap. The RMA had gone eight years without a group like EDS.'

During 1999, Gary Taylor, planner Denis Nugent and landscape architect Stephen Brown were jointly involved in cases before the Environment Court concerning the Waitākere Ranges. During breaks in the court proceedings, they would go out for coffee and lament the poor state of environmental advocacy in New Zealand. As Brown recalls,

> Gary and I had some coffees at Starbucks and he said 'what do you think about restarting EDS?' I knew relatively little about EDS then, so he went through the history and the people who had been involved. They included a lot of well-known resource management lawyers. I thought it was a great idea. Community groups were struggling to be heard in hearings and it was time to provide a counter-balance to the free-market view of resource management that was unfolding.

There was one case in particular that galvanised Gary Taylor to resurrect EDS. It concerned a wild, dune-backed, windswept beach stretching some 20 kilometres along the north-eastern coastline of the Auckland region. Very popular with surfers, swimmers and fishers, Pakiri comprised the longest undeveloped beach on Auckland's east coast. Its significance was reflected in provisions in the Rodney District Plan which sought to protect the natural and wild character of the beach.

Arrigato Investments Limited and Evensong Enterprises Limited owned land high up on Pakiri's southern headland. The companies proposed to subdivide the land into 14 rural-residential lots. To maximise views, the houses were to be built on the ridgeline, where they would be highly visible from the beach. As a form of 'offset' for the negative visual impacts of the buildings, the developers had offered to plant indigenous vegetation on the headland's steep coastal slopes. The land had long ago been cleared for farming and was now covered in weeds. Although re-establishment of indigenous bush on the headland would be an environmental improvement, there was some doubt as to whether the planting would succeed on such exposed terrain.

The Arrigato proposal required approval as a non-complying activity under the Rodney District Plan. Consent was initially declined by the Rodney District Council but in October 1999 its decision was overturned on appeal by the Environment Court (the renamed Planning Tribunal). This was despite strong opposition from the Auckland Regional Council which considered Pakiri to be 'a wild and scenic coastline of regional significance which should be protected'.

Gary Taylor thought it wrong that consent had been granted for such a proposal. As well as having a negative impact on one of Auckland's most notable undeveloped beaches, it set a dangerous precedent for future coastal development elsewhere. As Taylor commented at the time, 'there seems to be developing in the case law a free-for-all where district plans are routinely overridden by ad hoc applications.' In his mind, the case further highlighted the need for an environmental legal advocate.

In the same year, another case decided by the Court of Appeal (*Bayley v Manukau City Council*) set a further unhelpful legal precedent for the environment. The court created a legal fiction, assuming something to be true when it was in fact false, termed the

The development at Pakiri Beach was proposed for the southern headland (shown here near the top of the image), where houses were to be built along the ridgeline. Gary Taylor saw the decision to grant consent not only as damaging for the beach itself but as setting a dangerous precedent for future coastal development elsewhere. CRAIG POTTON

'permitted baseline'. This enabled decision-makers to disregard any potential adverse effects of a proposal, if the same effects could hypothetically be created by other activities permitted by the district or regional plan, even though they had not yet been established and might never occur. Given that structures were typically permitted in rural zones due to the practical necessity of enabling farming activity such as the construction of farm buildings, the application of the permitted baseline approach potentially opened up the floodgates to other (non-farming) structures such as houses, being built in rural areas.

It was also in 1999 that the National Party, which had been in power for nine years, was defeated by a coalition between the Labour Party (led by Helen Clark) and the Alliance (led by former Labour minister Jim Anderton). Jeanette Fitzsimons, co-leader of the Green Party, won the electoral seat of Coromandel. In only the second election under the new MMP (mixed-member proportional) voting system, this win, along with their 5.2 per cent of the votes, gave the Green Party seven seats in Parliament. The political tide was starting to turn against the strong neo-liberal economic policies that had been pursued by the previous Labour- and National-led governments.

—ooo—

EDS WAS FORMALLY reconstituted on 11 November 1999. The new set of directors included a mix of lawyers, planners, landscape architects and scientists. Charles Wurster, who had initially encouraged David Williams to establish EDS back in 1971, agreed to join the new board, as did previous long-serving EDS director Dick Bellamy. A new board member was Raewyn Peart, a former lawyer at Russell McVeagh, who had become involved with the

Maruia Society soon after its establishment in 1988. Peart subsequently moved to South Africa, but returned to New Zealand in 2001 and went on to lead EDS's environmental policy work.

EDS's rebirth was profiled in a *New Zealand Herald* article titled 'Environment watchdog returns to the fray', where journalist Philip English described the remit of the reformed entity.

> An environmental organisation which disbanded in 1988 after 18 years of fighting to keep New Zealand clean and green is being revived. EDS is relaunching itself as an environmental lobby group ready to go into courtroom battles against what is sees as unscrupulous development. Gary Taylor said that protection of New Zealand's landscapes would be a priority, although its advocacy would not be limited to landscape issues.

Quoting Gary Taylor, the article went on to explain:

> The RMA had generated an enormous amount of work for planners, lawyers and scientists ... 'but there is no effective public interest advocacy. It is all private interest advocacy. The whole system seems out of balance. An essential ingredient is missing.' Many groups were reluctant to take legal action against development proposals because of the complex issues involved, the costs and the risk of facing heavy court-imposed costs. But EDS would not be frightened off and would contest significant cases.
>
> One of the reasons for reviving the group was the diminished role of DOC [Department of Conservation] in conservation advocacy. Because of budget cuts it was now rarely involved in the planning process for development affecting the coastline or landscapes. Another reason for re-establishing EDS was to get younger people involved in environmental debates, as those who had been active in earlier years were getting longer in the tooth.

EDS had no staff, no office and little money. But this did not stop it becoming immediately active in environmental protection issues, drawing on Gary Taylor's voluntary commitment to the cause and pro bono support from a sympathetic network of lawyers and experts. EDS lived hand-to-mouth financially for many years.

—ooo—

EDS HAD ONLY JUST been re-established when it was once again embroiled in issues to do with the future of the remote Karikari Peninsula in Te Tai Tokerau Northland. In March 1999, then owner American businessman Paul Kelly applied to the Far North District Council for resource consents to build a country club development on the land where the Gold Coast-style resort had been proposed. It was to consist of some 384 accommodation units, a lodge and golf complex, and a vineyard. The council processed the application on a non-notified basis, meaning there was no opportunity for public submissions, and it ultimately granted the consents. The Department of Conservation only learnt about the multi-million dollar scheme when it saw the item on a council agenda.

After hearing about the development, EDS joined local iwi Ngāti Kahu to judicially review

During the late 1990s, the Karikari Peninsula (shown here looking along the dunes to the south-west) was the subject of a further development proposal that included 384 accommodation units, a lodge and golf complex and a vineyard. EDS successfully opposed the development. CRAIG POTTON

the council's decision not to notify the consents in the High Court. Then Russell McVeagh lawyer Christian Whata (now a High Court judge and Law Commissioner) agreed to act for EDS and the iwi on a pro bono basis.

As Gary Taylor recalls, 'this was my first introduction to how Russell McVeagh worked. The drafting of proceedings went right through the night. We went out for a break at around 3am and I was amazed to see more people in Fort Street and Queen Street than you would see during the daytime.' The proceedings were finally lodged in February 2000.

The district plan provisions were weak, which meant that it would be difficult to stop the development entirely. After revisiting the site, Taylor concluded that the most that could likely be achieved was to reduce the scale of the development and keep it away from the beach. Negotiations with Paul Kelly ensued and they resulted in an agreement which achieved this. Kelly also committed to not seeking any future expansion of the consented accommodation and to consult in good faith with EDS and Ngāti Kahu on any future development of the site.

The development subsequently went ahead and 14 villas, as well as a modest lodge with 10

guest rooms, were built alongside a golf course and vineyard. It was a significant reduction from the 384 units originally sought, with the buildings all set well back from the wetland and the beach.

But that was not to be the last attempt to develop the beach. Without any public notification, and without notifying EDS or Ngāti Kahu, Kelly applied for consent to create a residential subdivision for 12 houses overlooking the beach. This was granted despite the efforts of Ngāti Kahu to stop it through the courts. In 2013, the property was sold to Chinese investors and the new owners have indicated interest in further developing the site for high-end tourism. No further applications had been lodged by the time this book went to print but EDS is still keeping a watching brief on developments.

The Karikari experience provided a vital early lesson for EDS. If it wanted to prevent future development of a site, restrictions needed to be secured through legal covenants over the land.

—ooo—

WHEN THE GREEN PARTY gained seats in Parliament after the 1999 election, it was not part of the formal governing coalition between Labour and the Alliance. However, the Greens agreed to support the government on confidence and supply in return for input into the Budget and legislation. In 2000, the Greens negotiated a $15 million 'green'

EDS and Ngāti Kahu managed to negotiate a reduction in the size of the development proposed for the Karikari Peninsula from 384 accommodation units to just 14 villas and a modest lodge associated with a golf course and vineyard. The completed development is shown here in 2008. CRAIG POTTON

For a decade, until 2013, the EDS team travelled around the country to run workshops and provide legal advice to local communities. Shown here is the EDS team discussing environmental issues with local residents at Wainui Beach, Gisborne, before a workshop. EDS former in-house lawyer Nicola de Wit is third from left, EDS Chief Executive Gary Taylor fourth from left and EDS legal researcher and policy analyst Kate Storer second right. RAEWYN PEART

package in the first Labour/Alliance Budget. Importantly, this included legal aid and other assistance for environmental organisations. The availability of this funding was a major help in re-establishing EDS; both through helping to fund litigation and providing financial support for the society to provide community advice and education.

In 2001, EDS director and lawyer Raewyn Peart returned to Auckland from South Africa. With a four-year-old daughter in tow, she was reluctant to work in a full-time professional job, and was looking for part-time work that would fit around parenting duties. Taylor suggested that she apply, on behalf of EDS, to the newly established Community Environment Fund for support to write a community guide to the RMA. This would mirror its *Conservation in New Zealand: A Citizen's Guide to the Law* that EDS had published in 1985. Having recently written a website-based state of the environment report for the Durban City Council, Peart was well aware of the rising popularity of the internet, and she thought the guide would best be located on a website.

The application to develop a web based guide to the RMA was duly made to the Ministry for the Environment in 2002, and after some delay a response came back. The ministry somewhat grudgingly agreed to provide the funding with one proviso. It was not convinced there was sufficient demand from community groups to access information from the internet and therefore it required EDS to undertake a survey of community groups to establish such demand before it could access the funding for the project.

As Peart later observed, 'in hindsight the request was quite outmoded. With the rapid growth and domination of the internet as a way of accessing information, it was clearly the way to go. I had been working for a research organisation in South Africa where the internet was commonplace. EDS was just ahead of the times in New Zealand.'

Fortunately, the survey showed that community groups were using the internet and the project went ahead. To help reach a broader audience, the material was also produced as a hard-copy guide in 2004. This led to the production of a whole series of community guides by EDS on topics as diverse as landscape protection, coastal development, freshwater management, biodiversity protection and oceans management. The hard-copy community guide to the RMA went to three editions. Many of the guides were written by Raewyn Peart, but others also contributed. Coastal planner and writer Lucy Brake helped out in later years and was the prime author of the last three guides on the coast, biodiversity and oceans.

Almost 20 years later the community guide website is still active. It is now hosted by the Environment Foundation and has been expanded to include a wide range of environmental topics.

The website is widely used by community groups, students, local government politicians and others and has recently received funding from the Ministry for the Environment to undergo a major update to include the latest resource management reforms.

For around a decade, from 2003 to 2013, the Community Environment Fund also supported EDS to provide a nationwide RMA community advice service and to run community workshops around the country. Along with the community guides, this effort was designed to upskill community groups and individuals in effectively engaging in the RMA process.

The workshops proved very popular with local communities and were held in a wide range of places, including Kerikeri, Whitianga, Kawhia, Taupō, Tūranga-nui-a-Kiwa Gisborne and Whakatū Nelson. They provided both general information about the RMA and engaged with local environmental issues. At some workshops, Environment Court judges attended and gave advice on how to present an effective case in front of the Environment Court. The workshops also profiled how to access official information.

In July 2013, the Ministry for the Environment decided to stop funding EDS's community advice service and workshops. EDS was unable to find a replacement supporter and the service has never been re-established.

—ooo—

MINING ON Te Tara-o-te-Ika Coromandel Peninsula was still a live issue when the Thames-Coromandel District Council notified a proposed new district plan in 1997. The council had continued its hard line against mining, so in this first proposed plan under the RMA, it had identified mining as a 'prohibited activity' (a category not available under the former Town and Country Planning Act) in sensitive areas including the conservation and coastal zones. This meant that an application could not be made for mining within those areas unless the plan was changed. In all other areas, mining was a non-complying activity, which meant that applications for the activity could be made but they needed to overcome

During the 2000s, the EDS team assisted many communities experiencing coastal development pressures, including at Waipuka Ocean Beach, Heretaunga Hastings shown here. EDS Chief Executive Gary Taylor (on the left) can be seen discussing the issue with a concerned local along with then in-house lawyer Shay Schlaepfer (in the centre). RAEWYN PEART

additional hurdles before consent could be granted. The area where mining was proposed to be prohibited included large areas of the 'Hauraki goldfields' where significant gold and silver resources were located.

The New Zealand Minerals Industry Association, which represents the mining industry, was not surprisingly opposed to this restrictive approach. In July 2004, after the council declined to accommodate the mining industry's concerns in its decisions on submissions, the association appealed to the Environment Court. The court found in the association's favour and changed the status of mining in the coastal zone from prohibited to non-complying. The court concluded that the RMA was based on a 'permissive, effects-based philosophy' and 'unless it can be definitively said that in no circumstances should mining [the activity in question in this case] ever be allowed on a given piece of land, a prohibited status is an inappropriate planning tool'.

Coromandel Watchdog, which had been a submitter on the proposed plan, then appealed to the High Court, as did the council. The Ministry of Economic Development became a party in support of the mining industry. After hearing all the submissions, the High Court adopted a similar narrow legalistic approach

Mining continues to be a controversial issue on the Coromandel Peninsula, as evidenced by this protest sign. EDS assisted Coromandel Watchdog to uphold the application of prohibited activity status to mining applications in the coastal zone of the peninsula.
RAEWYN PEART

to that of the Environment Court and, in its decision issued in December 2005, upheld that court's findings. At this point the Thames-Coromandel District Council threw in the towel and decided not to appeal further.

The court decision placed an exceedingly restrictive requirement on the use of prohibited activity status. It effectively meant that the category could virtually never be used. As a result, district plans would be unable to head off inappropriate activities or developments from the outset. Everything would be up for grabs. This further entrenched the permissive approach to applying the RMA and, as far as EDS was concerned, was a dangerous precedent which could not be left to lie.

EDS was not a party to the district plan proceedings as the society had not been operational when the plan was notified back in 1997. It therefore did not have standing to lodge an appeal. So the society put its support behind Coromandel Watchdog to seek leave to appeal to the Court of Appeal. Auckland lawyer Rob Enright, who was at that time a partner at Kensington Swan, acted for Watchdog.

In the first instance, leave was sought from the High Court to appeal further, and this was declined. The court was heavily influenced by the fact that the Thames-Coromandel District

Council no longer wanted to be involved in the proceedings and no other councils had shown interest in the matter. Then Coromandel Watchdog sought special leave from the Court of Appeal itself.

Fortunately, the appeal court grasped the significance of the issue and agreed to hear the case. By this time Auckland City Council and Auckland Regional Council had been alerted to the proceedings. They sought leave to become parties due to the broader implications of the decision for other councils around the country.

After hearing legal argument from the various parties, the Court of Appeal concluded that the lower courts' interpretations were too restrictive and wrong. This meant that the prohibited activity status remained a category that councils were able to use in a wide range of circumstances. The overly restrictive approach that had been adopted by the Environment Court had been beaten back, at least in this one area, and the wider utility of prohibited activity status in plan-making was confirmed.

ANOTHER AREA that EDS became involved in soon after its re-establishment was the challenging issue of how greenhouse gas emissions should be managed. There had been growing public awareness of the threat of global warming over the previous decade. New Zealand had signed up to the United Nations Framework Convention on Climate Change in 1992, alongside more than 180 other states, and the convention came into force in 1994. In 2002, Parliament subsequently ratified the Kyoto Protocol, which committed New Zealand to reducing its greenhouse gas emissions during the first commitment period of 2008–2012 to 1990 levels.

During the early 2000s, there had been renewed interest in electricity generation with proposals to establish combined cycle gas-fired power plants at Ōtāhuhu in Auckland and at Huntly, and to expand the plant at Stratford in Taranaki. All three proposals claimed that their new power station would reduce overall national greenhouse gas emissions. This was on the basis that it would replace generation from existing installations which were less efficient and had higher greenhouse emissions.

Garry Law, a civil engineer specialising in water engineering, became involved in the issue on behalf of EDS. He had come to know Gary Taylor through their joint involvement in local body infrastructure provision. Law had been invited onto the EDS board when it was reconstituted in 1999 and he became the society's climate change expert.

Law had done some calculations and he found that the cumulative emissions of the three proposed new plants was far in excess of the amount being produced by less efficient thermal plants. In short, if all three plants went ahead, national greenhouse gas emissions would increase even if all the inefficient plants were shut down. As Law explained, 'if they are all built together they have the potential to emit 3.8 million tonnes of CO_2 a year into the atmosphere which is a 5.3% blowout on the Kyoto Protocol target of getting emissions back to the 1990 level'.

So EDS became involved in the consenting processes with a focus on the need to mitigate greenhouse gas emissions emitted from burning gas. Barrister Richard Brabant ran the Ōtāhuhu and Huntly cases for EDS and Law gave expert evidence.

It turned out to be an uphill battle. EDS asked Environment Minister Marian Hobbs to call in the three applications so they could

Directors' breakfast at the 2019 Climate Change and Business Conference. EDS

be considered together, and conditions relating to the closure of older plants and carbon offsets such as forest plantings jointly addressed. Because the forest offsets might be within a different council jurisdiction than the proposed power plant, it could be difficult to impose offsetting conditions if each consent was determined by a single council. However, Minister Hobbs declined to take any action.

EDS went to the Environment Court, but in the end all three power plants were consented without offsetting conditions being imposed. Although the court acknowledged that there would be greenhouse gas emission impacts, it concluded that addressing them was outside its jurisdiction. In the end, only two power stations were ultimately built, the one at Ōtāhuhu did not go ahead. They were all competing against each other for a limited gas supply and there wasn't enough fuel to service them all.

Although EDS's efforts failed to achieve mitigation measures on the power plants, the cases were important in other ways. As EDS director and law professor Barry Barton observed, 'they were important markers to push our understanding forward on how the RMA worked in respect of climate change. It may not seem like it from the outset, but it was significant that the judge agreed that carbon dioxide was a pollutant in terms of Section 15 of the RMA. We were pushing the limits on the RMA to see how far it would go.'

In 2004 the Resource Management (Energy and Climate Change) Amendment Act was

Garry Law obtained a Bachelor of Engineering degree from the University of Canterbury and he went on to become a civil engineer specialising in water engineering. Law worked for the Auckland Regional Council for many years, becoming its chief engineer and then general manager of operations. Between 1991 and 1994 he was the chief executive of Watercare Services, followed by two years managing Brisbane Water. Law is also an experienced archaeologist and an honorary life member of the New Zealand Archaeological Association.

Law served as an EDS director for 20 years, between 1999 and 2019. He chaired the society's finance committee for many years. Law was EDS's first climate change specialist and he put together a regular climate change newsletter for EDS, as well as weekly blogs. He also attended climate change convention meetings in The Hague and Marrakesh on behalf of EDS. Law was EDS's first webmaster and he created the web framework for the first EDS online guide to the RMA.

passed. This excluded regional councils from considering the impact of greenhouse emissions on climate change when creating rules in plans and considering discharge permits. This meant that such emissions could not be considered in any future consents for power plants or other greenhouse gas emitting facilities. The ostensible reason for the exclusion was that emissions would be addressed at a national level by a carbon tax that the government was proposing to introduce at the same time.

It was around this time that EDS decided to convene a climate change and business conference. Elizabeth Edmonds, who was involved in organising many of the conferences, describes their inception:

> Jo Hume had been running the New Zealand Business Council for Sustainable Development so knew EDS quite well. She then moved back to Australia. Jo and Gary were catching up by phone one day. Gary was saying how pathetic the efforts of Australia were on climate change to which Jo responded, 'no actually we're doing a fair bit over here'. She proceeded to rattle off all the things that Australia was doing. 'But', she went on, 'New Zealand is pathetic, it's not doing anything'. Gary says 'no' and starts defending New Zealand. Then he says 'this is ridiculous, the two of us don't even know what's going on across the ditch. Wouldn't it be great if there was a forum where business and government and other interested parties from both countries could come together on climate change issues to learn from one another.' So they decided to create it.

The first conference was held in Auckland in November 2004. It was organised by EDS in partnership with a raft of business entities including the New Zealand Business Council for Sustainable Development, the New Zealand Council of Trade Unions, the Business Council of Australia, the Australian Business Council for Sustainable Development, the Australian Council of Trade Unions and the US Pew Center on Global Climate Change.

Given the considerable financial risks involved in running such a large event across two countries, EDS established a separate corporate entity called the Australia-New Zealand Climate Change & Business Centre to convene the conferences.

It turned out to be a very successful event, with 220 attendees. Around a quarter had travelled from Australia. As Elizabeth Edmonds recalls,

> we had a lot of major emitters talking at the conference and some criticised us for that. Rio Tinto was one of our sponsors. But I always believed, if you don't have everyone in the room you are just preaching to the converted. Everyone walked away from the first conference thinking it was hugely valuable to have a forum where all parties can come together to learn and try to figure out what is going on. The conference reputation grew and the event evolved into a serious forum for government and business, in particular, to hear from one another.

The second conference was held in Adelaide in February 2006. Jo Hume was now pregnant so Elizabeth Edmonds took over the organising role. By that time the European Union had established the world's first major carbon market with an emissions trading system. But things were not so rosy in New Zealand where, in 2005, the Labour-led government had been forced to abandon the carbon tax proposal when it lost support from coalition partners United Future and New Zealand First.

Although the government was slow to take action on climate change, businesses were becoming engaged with the issue. In particular, a group of major corporates formed the Australian Business Roundtable on Climate Change. They did some research which showed that the cost of inaction on climate change would far exceed the cost of taking action. EDS saw it as a major coup when all the major companies used the 2006 Adelaide conference to sign off the report.

EDS initially planned to hold the conference every 18 months, but after interest rapidly increased in climate change matters, it convened the event every year. Edmonds has a particular memory from organising the climate change and business conferences with Gary Taylor.

> There was an organisation called the Climate Institute that rapidly appeared out of nowhere on the back of a large donation from a family trust. Gary was in Australia to do the annual conference marketing trip with me, looking for funding and sponsorship. We also chatted to people about what should be in the programme. So we decided we should talk to these new guys on the block.
>
> We brought a summary of what the conference was about to the meeting to tell them what we were doing and explore how we might work together. The guy just sat there and flipped through the presentation without even looking at it.
>
> I was trying to win this guy over, but ten minutes into the conversation Gary

Both New Zealand Prime Minister Helen Clark and Australian Prime Minister Kevin Rudd addressed the 2008 climate change and business conference. They are shown here in a dialogue facilitated by business journalist Rod Oram (centre). EDS

> cut me off. He looked straight at the guy, said 'you're clearly not interested in a word we're saying, and if you're not interested and have no time for us, I don't have time for you either'. Gary then got up and walked out of the room. I have never forgotten that. Gary was absolutely right. He called it for what it was.

Interest in the conference was heightened in 2007 when, in Australia, the Rudd Labor government proposed an emissions trading scheme called the Carbon Pollution Reduction Scheme. A year later, the New Zealand Labour-led government created the New Zealand Emissions Trading Scheme with the passage of the Climate Change Response (Emissions Trading) Amendment Act 2008. The Act came into force on 25 September 2008, just days before the 8 November general election.

From 2005 to 2008, David Parker was Climate Change Minister in the Helen Clark government. He put together a Climate Change Leadership Forum, chaired by notable businessman and philanthropist Sir Stephen Tindall, to provide advice on policy issues relating to climate change and emissions trading. Gary Taylor was appointed to the group. As Parker observed, 'It was a coming together of environmental groups, energy companies and corporate interests who were interested in doing something about climate. It was a remarkable group, and to me it demonstrated a change point for EDS, when it started to work with development interests to get good outcomes. EDS was trying to work with people rather than in opposition to them.'

Prime Minister Helen Clark addressed the 2008 conference in Auckland, as did newly elected Australian Prime Minister Kevin Rudd.

COURTESY OF ELIZABETH EDMONDS

Elizabeth Edmonds is an American citizen based in Sydney. She holds a Bachelor of Science degree from Stanford University and Masters degrees in Public Policy and in Politics and Research, from Macquarie University in New South Wales. Edmonds organised the climate change and business conference for EDS from 2006 until the business was sold in 2011. When she started working on the climate change and business conference, Edmonds had just left a job as fundraiser for Oxfam.

'When I started working for EDS, Gary met me in Brisbane for the first round of conference meetings. We had a very good meeting with Rio Tinto which confirmed its sponsorship. Gary then said "time for lunch". I replied, "there's a little takeaway over there", but he was looking for something else. "This looks like a good fish restaurant," he said and disappeared upstairs. It was a nice restaurant overlooking the river. I was like, "oh, this is a bit different to Oxfam". Gary then explained, "if you're going to get the best out of people you need to treat them like they are the best". I thought, "I think I'm going to like working here". That was the difference'.

As Elizabeth Edmonds recalls, 'at that point we thought it was all going to happen and there would be great policy action on climate change. Australia and New Zealand thought they could work together and we had politicians going back and forward to look at how to price carbon and trade credits. The conference provided an important venue for some of those key discussions.'

But then the prospect of government action started to fade and the benefit of maintaining cross-Tasman dialogue receded. In Australia, the Rudd government struggled to get the Senate numbers to pass legislation for the Carbon Pollution Reduction Scheme. It was finally passed in 2011 but was then subsequently repealed on the change of government.

Things weren't much better in New Zealand. The Labour Party lost the 2008 election and a National-led government took over the Treasury benches. Although it didn't repeal the emissions trading scheme, the new government severely weakened it in various ways.

Most notably, changes were made to allow the unlimited importation of emission units from overseas. This served to crash the price of carbon credits to just $2 a tonne in 2013. This meant there was no effective price signal incentivising a reduction in emissions and councils were still unable to scrutinise greenhouse emissions from developments consented under the RMA. This weakening of the emissions trading scheme coupled with the inability of councils to consider emissions in consenting created a large policy gap when it came to climate change.

Seven climate change and business conferences had been held when EDS sold the

The climate change and business conference, which EDS restarted in 2015, has gone from strength to strength. Shown here is a panel during the 2022 conference facilitated by broadcaster Kathryn Ryan (far left) with (moving right) Secretary for the Environment Vicky Robertson, Climate Change Commission Chair Rod Carr and Auckland City Mayor Phil Goff. DAVID HARTMAN

business to an Australian competitor after the August 2011 conference. The sale provided badly needed funds to the society. By that time it had been held in Auckland (twice), Adelaide, Brisbane, Melbourne, Sydney and Wellington. At its peak (in 2009) there were 525 people in attendance.

The conferences had substantial political influence. As observed by Nick Smith, former Minister for Climate Change Issues in the Fifth National Government,

> you cannot have a discussion on climate change without engaging with business. Too often we have seen countries completely fail on climate change because the business and environmental communities never connect. The Climate Change and Business conferences have been pivotal for making progress on the horrendously difficult challenge of getting emissions down.
>
> It was through the conferences that I, as Climate Change Minister for New Zealand, did the deal with the Australian government to work towards developing an integrated Australia–New Zealand emissions trading scheme. It would have made great sense given the degree the economies of New Zealand and Australia are integrated. One of my big disappointments was that the Rudd government lost its nerve and walked away from it at the last moment.
>
> The most fragile Cabinet meeting, in my more than a decade in the Cabinet room, was holding the line in 2010 on New Zealand proceeding with the emissions trading scheme. It was the contacts EDS had built with business that enabled myself and the [John] Key Government to withstand the huge pressure to follow Australia in walking away from the ETS.

When EDS sold the climate change and business conference it was on a deferred payment schedule. The purchaser failed to make the final payment, which was due in March 2014, and so the business legally reverted back to EDS. The society resumed running the climate change and business conference the following year, this time solely as a New Zealand event. EDS events director, Fiona Driver, helped organise the events in association with Gary Taylor and former EDS lawyer Madeleine Wright. Six conferences were held in the series before the 2021 conference was deferred due to Covid-19 restrictions. It was eventually held in 2022 followed by another in 2023.

The conference had re-started small, but after five years there were a regular 400 plus people in attendance. Reflecting the recent upsurge in interest in climate change matters, the August 2022 conference, held in partnership with the Sustainable Business Council and the Climate Leaders Coalition, had over 700 registrants. Looking back at the conferences since 2015, Driver noted,

> I am really proud of the leaps and bounds made with the climate change conference. It enabled us to build relationships with businesses such as Westpac and IAG which EDS would not normally have exposure to. That the conference was well respected was shown by sponsors coming back year after year. It was never a hard sell. I think people always felt they got a lot out of the conferences which is why they were prepared to keep coming back.

The conferences have also helped pave the way for more rigorous climate change policy in the form of the Climate Change Response (Zero Carbon) Amendment Act 2019. As Wright observed, 'one of the focus areas of the conferences was getting climate change legislation in New Zealand similar to the UK model. After several years of working on the conference, and also on the side with other organisations, it became a reality which was fantastic. It was a huge highlight in my life to have a role in that and in bringing the business community along in support.'

The Amendment Act represented a major turning point in New Zealand's climate change policy. It established an emissions reduction target, emission budgets to create stepping stones towards meeting the target, and an independent Climate Change Commission to keep the government honest. The emissions trading scheme was also strengthened with the price of carbon credits rebounding. And, in 2020, the provisions preventing councils from considering the impacts of greenhouse gas emissions in consenting were repealed by the Resource Management Amendment Act 2020. This will potentially allow offsetting measures to be considered once again.

10. Protecting the coast ... again

WHEN EDS WAS RE-ESTABLISHED in the late 1990s, one of the urgent issues was coastal development. The country was experiencing a coastal development 'gold rush' with exponential growth in subdivision and development of holiday and retirement homes along the coast and lakesides. Beach after beach was being developed as an ever-expanding ribbon of urban development was wrapping itself around the coastline.

AS GARY TAYLOR observed at the time, 'coastal development on the Coromandel, in Northland and other places is burgeoning out of control'. He went on to posit the cause:

> the problem is the Resource Management Act, which works reasonably well for most places, but isn't up to the task of protecting the most beautiful parts of our countryside because the pressures are just too great to resist. Development and subdivision are occurring everywhere without regard to the environmental effects of such development. We are losing our wild places and the built environment is taking over everywhere.

A key driver was baby-boomers who, having reached the stage of their lives when they were financially secure, were looking to purchase a holiday retreat for themselves and their families. And sea views and proximity to the water were paramount.

Where there was demand, developers were lining up to meet it, and there was no shortage of rural landowners willing to sell as the removal of agricultural subsidies during the 1980s rendered many coastal farms uneconomic.

The regulatory framework was poorly configured to deal with such pressures. Government had promulgated a New Zealand Coastal Policy Statement in 1994, but it was high-level, somewhat vague and had been largely ignored by councils and the courts when it came to coastal subdivision. This meant it was difficult to stop the carving up of coastal properties through the legal process, so EDS looked to other ways of achieving better environmental outcomes, while also seeking to strengthen the law.

EDS WAS INSTRUMENTAL in heading off early development proposals for Waikawau Bay (see chapter 7), and the land behind the beachfront had been incorporated into a Department of Conservation reserve. But that left the sensitive northern headland still in private ownership.

During the mid-1990s, the headland was acquired by American businessman and Karikari property developer Paul Kelly. Kelly eventually decided to gift the 149-hectare headland farm to the University of Auckland. This was on the condition that proceeds from its sale would be used to support the university's proposed new business school. The donation was to be matched by a 1:1 subsidy from the government's Partnerships for Excellence scheme. The university put the land on the market, with a minimum price tag of $3.5 million, and indicated an intention to sell to the highest bidder.

Waikawau Bay was one of the last undeveloped beaches on the east coast of Te Tara-o-te-Ika Coromandel Peninsula, and as far as EDS was concerned, the land needed to be secured in public ownership to extend the Department of Conservation reserve.

Gary Taylor thought that protecting the headland was particularly important, as it shaped and enclosed the coastal experience of the beach. So EDS mounted a three-pronged campaign. The first prong consisted of scaring off potential buyers, the second sought to convince the university to donate the land to the Department of Conservation, and, failing that,

---

Previous pages: In 2003 the north headland of Waikawau Bay was put on the market by the University of Auckland after it was gifted the land by American businessman Paul Kelly. EDS again swung into action to help secure the land in public ownership. RAEWYN PEART

After helicoptering Conservation Minister Chris Carter down to Waikawau Bay to experience the beauty of the wild undeveloped beach for himself, EDS Chief Executive Gary Taylor was able to obtain a commitment from him that the government would buy the northern headland property, which it did. The whole bay is now secured in public ownership. RAEWYN PEART

the third sought to persuade the government to buy the land.

In a strongly worded media statement in January 2003 EDS sought to dissuade any potential buyers of the headland by announcing that the society would vigorously oppose any plans to carve up the property. An interested party sounded out EDS and the Department of Conservation on a possible development involving 20 houses, but was told in no uncertain terms that both would strongly oppose any such development. This served to dampen buyer interest in the property.

Taylor also met with John Hood, who was University of Auckland Vice Chancellor at the time, seeking to persuade him that the property should go into public ownership. This was supported by Conservation Minister Chris Carter and Thames-Coromandel Mayor Chris Lux, who both argued that the university should either hand the land over to the Department of Conservation outright, or sell it at a reduced price. Even Prime Minister Helen Clark added her voice to the cause. But Hood remained staunch. As far as he was concerned, the land would be sold to the highest bidder.

Taylor then discovered that the Department of Conservation had been given a previous opportunity to buy the land at a lower price, but it had declined because the headland lacked sufficient ecological or conservation value. Persuading the government to front up with the money to buy the property now, at a higher price, was not going to be easy.

Fortunately, Taylor had a personal connection with the Conservation Minister Chris Carter, who had previously been his next-door neighbour. Taylor contacted Carter and invited him to 'come down and have a look at the property'. The minister was about to head overseas so time was of the essence. A helicopter ride was the only practical way to get him to Waikawau Bay and back before he left on his trip. As Taylor recounts, 'I tried to get the Natural Heritage Fund to organise a helicopter to fly Chris down, but in the end they wouldn't do it. So I put the helicopter on my credit card. I got hold of TVNZ and the *New Zealand Herald* and they all piled into the helicopter with me and Chris.'

> It was a very nice Coromandel day. After we landed on the beach I said to Chris 'you need to buy this'. The cameras were rolling and he said 'yes, we are going to buy this'. We had presented Chris with a fait accompli, and he had little time to consider the matter, as he was heading away on an overseas trip the next day. But I don't think he minded being put in that position as he was a good Conservation Minister.

In February 2003, the government formally announced that it would purchase the property. It paid $3.45 million, just under the asking price. The entire beach was now secured for the benefit of the public and the business school also got its funding. Taylor found the outcome particularly satisfying. 'It was one of the best direct specific engagements I have had with a cabinet minister. It was short, sweet and had a good outcome very much in the public interest. The environment on the north end of the beach is now protected in perpetuity.'

AFTER THE ECONOMIC reforms of the late 1980s and 1990s, when rural subsidies were withdrawn, many farms in Pēwhairangi Bay of Islands were proving unprofitable and were becoming increasingly run down. At the same time, land values were rising, buoyed up by the coastal development boom, as were council rates. Farming families were looking for a way out of the financial squeeze and started selling their land to coastal developers.

Some developers sought to maximise their profit by carving up the farms into as many sections as possible. But others wanted to do things differently. These were generally well-heeled business people who wanted to create high-quality, low-intensity developments that provided a financial return, but also contributed to restoring the degraded coastal landscape.

One of these developers was Peter Cooper. Northland-born, Cooper was a former Russell McVeagh lawyer who had gone on to create a successful property and investment company in the United States. Although he now spent much of his time there, Cooper was keen to own a large piece of coastal land in his home country. After five years of searching for the right property, he heard about a farm on the market on the Purerua Peninsula, a thick wedge of land on the north-west edge of the Bay of Islands. It was a property with significant cultural as well as natural values, having been the location of much early Māori settlement as well as some of the earliest contacts between Māori and Europeans.

A property on the Purerua Peninsula in Pēwhairangi Bay of Islands, known as Mountain Landing, was the subject of one of the new models of low-intensity coastal development which emerged during the 2000s. Houses were sensitively located off ridgelines, native vegetation and wetlands were restored and future subdivision of the site prevented through legal covenants placed on the land titles. CRAIG POTTON

At that time, the property was owned by the Mountain family, who had farmed it for several generations. But by the 1990s, they were struggling to make an income off the land. Stock was largely left to roam and, unable to afford fencing, they had planted gorse along the coastal clifftops to stop stock falling over the edge. In the end, they decided to put the property on the market and, in 2000, Cooper bought the 383-hectare farm for $8 million.

After spending time getting to know the land, Cooper went on to design a subdivision which included 39 house sites and a 262-hectare working farm. Extensive areas were to be replanted in indigenous vegetation, wetlands were to be restored, and 12 hectares of heritage areas protected. Provision was made for local hapū members and the public to access

The Mountain Landing development, which EDS supported, took a very rundown farm and did extensive replanting and restoration works to revive the natural systems on the land. CRAIG POTTON

the coast, although the bulk of the property was to be off-limits as a gated community. Importantly, future subdivision of the land was to be prevented through covenants placed on the property titles.

EDS was keen to support 'good' coastal development and Cooper's vision for a sensitively designed, restorative development struck a chord. It seemed a positive step forward, and a major improvement on the traditional intensive, grid-like subdivisions that bedevilled much of the country's coastline.

EDS worked with Cooper to improve the development by locating house sites away from visually sensitive areas. To achieve this, Gary Taylor brought in landscape architect Stephen Brown to advise on the location of the lots, and several were moved on his advice, to perch below ridgelines and be screened by plantings.

EDS worked in a similar manner with New Zealand businessman Craig Heatley, who

had acquired a 747-hectare Bay of Islands property called Bentzen Farm. A rural-residential subdivision over 142 hectares of the land was later named Ōmarino. It contains just 17 house sites. Stephen Brown again provided advice to the developer, on behalf of EDS, to help reduce the visual impact of the subsequent buildings.

Stephen Brown recalls that

> before Craig Heatley became involved there had been an attempt to develop the land. There were scars everywhere where someone had driven a bulldozer trying to create farms tracks and put in retaining walls and that sort of stuff. Our discussion focused on how to restore the property, and sensitively locate development within it to assist with restoration, rather than plonking as many houses as you could all over it. We made suggestions about relocating building sites and reducing the sizes of dwellings. I think it was quite successful in the end.

EDS worked with businessman Craig Heatley to reduce the visual impact of houses which were to built as part of a low-intensity rural-residential development on Bentzen Farm. Shown here is the EDS team visiting the site with then in-house lawyer Shay Schlaepfer (second from left). To her right are landscape architect Stephen Brown, developer Craig Heatley and EDS Chief Executive Gary Taylor. RAEWYN PEART

EDS came under some criticism for working with developers, particularly when it endeavoured to set up a 'Green Tick' scheme to acknowledge best-practice coastal development. But the society was adopting a pragmatic approach, seeking to achieve environmental gains where it could. Government had shown no interest in acquiring sensitive coastal land and the legal framework for the protection of coastal values was weak. Within this context, it seemed better to work with benign developers to lock in low-intensity development with associated restoration benefits rather than leaving the land vulnerable to poorly designed intensive development which was 'business as usual' around the coast.

RAEWYN PEART

Stephen Brown completed a Bachelor of Town Planning from the University of Auckland in 1978 and a Postgraduate Diploma in Landscape Architecture from Lincoln University in 1981. He worked for the Auckland Regional Authority for three years, undertaking the first assessment of the region's landscapes. In 1984 Brown moved into private practice as a landscape architect specialising in landscape assessment and planning.

Brown has undertaken landscape assessments of many parts of New Zealand including the Tāmaki Makaurau Auckland, Waikato and Te Mata-a-Māui Hawke's Bay regions and Manawatū, Tararua, Rangitikei, Ōtorohanga, Thames-Coromandel and Whangārei Districts. He is currently a fellow and past president of the New Zealand Institute of Landscape Architects.

Brown was a director of EDS from 1999 to 2011. He has provided expert landscape assistance to the society on many projects including Plan Change 13 in the Mackenzie Basin, the King Salmon case in the Marlborough Sounds, New Chums Beach, Coromandel and developments in the Bay of Islands.

—OOO—

WIDESPREAD PUBLIC ACCESS to beaches and rivers is a long-held New Zealand ideal. It is represented by the concept of the 'Queen's chain', a one-chain (20-metre) wide strip of public land running alongside water bodies. However, the Queen's chain is an ideal rather than a reality along many areas of the coast.

When land was cut up during early European settlement, such a strip was created in some parts of the country, often through the provision of a surveyed but unbuilt 'paper' road running along the coast. But in other parts of the country, land was privatised right up to the high-tide mark.

Recognising the public imperative to increase coastal access, the RMA requires the provision of a 20-metre-wide esplanade reserve or strip along the coast and waterways whenever adjoining land is subdivided into lots of less than 4 hectares. This is effectively when rural land is urbanised. However, the Act also gives councils discretion to reduce or waive this requirement in individual cases.

The issue of public access to the coast came to a head when G. L. Wolfe sought to subdivide the former campground at Te Puia Hot Water Beach into seven house lots. The campground adjoined the Taiwawe Stream (which flows into the beach), and the house lots were less than 4 hectares, bringing the esplanade reserve requirement into play. However, Wolfe was not happy to give away the full 20-metre strip of land, and he sought to reduce the width to 7 metres along some parts of his property. He argued that the 7-metre-wide strip still provided adequate public access and so the reduction from 20 metres should be authorised.

Both the Thames-Coromandel District

When the campground at Hot Water Beach was subdivided, the developer sought to reduce the width of the esplanade reserve. EDS supported the Tairua Environment Society to successfully argue for the full 20-metre provision. As well as ensuring ample public access along the edge of the development (which can be seen here as the grassed area in front of the new houses), the case also set a valuable precedent. RAEWYN PEART

Council and Department of Conservation caved in and agreed to the proposition. The council granted consent to the subdivision in October 2003. This left it to environmental groups to fight for the public's right to have the full Queen's chain. EDS put its resources behind the Tairua Environment Society to appeal the matter to the Environment Court. Ian Cowper provided legal representation on behalf of EDS.

At the hearing Wolfe argued for a trade-off; he would replant the riparian area and provide a public walkway in exchange for a reduction in the size of the public reserve. The Environment Court was not persuaded, noting that the long-term viability of the planting could not be assured and stating 'there needed to be compelling and robust reasons to depart from the requirement for the 20-metre esplanade reserve'. The development went ahead but with the full-width reserve.

The case set a high legal bar which has helped ensure greater public access to the coast and waterways since that time.

Ōhiwa spit is subject to cycles of erosion and accretion. When property titles came out from under the sea during the 2000s, an owner decided to sell the land and the purchaser sought to build a holiday home on it.
ROB SUISTED/NATURESPIC

—000—

ON 8 DECEMBER 2005, the *Property Press* ran an advertisement for the sale of a residential section at 2 Ohiwa Beach Road, Ōhiwa. In May 2006, Wellingtonian Jeff Richter bought the property for $180,000 and then sought consent from the Ōpotiki District Council to build a holiday home on it.

But this was no ordinary piece of land. It was located on the Ōhiwa spit, a small limb of sand on the eastern fringes of Ōhiwa Harbour, and a place which had an unusual history of development.

A ferry service used to operate across the Ōhiwa Harbour and the first buildings on the spit comprised a hotel, built during the 1870s to service ferry passengers, followed by a post office. Two years later, 36 sections were surveyed to form the Ōhiwa township.

Only a few sections were subsequently built on, but a store, ferryman's cottage and stables for coach horses were established. Unbeknown to the new settlers of the area, the land that they had built on was not stable, but subject to cycles of erosion and accretion. This had not been immediately obvious because the development had taken place during a relatively stable period for the spit. But things soon changed. After 1900, erosion suddenly accelerated, and within 15 years it was threatening the township. Buildings were hastily removed and the township abandoned.

It took around a generation for collective memories of the erosion threat to dim, and during the 1930s a few baches started to appear on the spit. However, severe storms soon eroded more land and the main street of Ōhiwa township ended up in the middle of a tidal channel. The baches were removed.

By 1945, the rate of erosion had slowed again, giving false hope that what was left of the spit was stable. The Department of Lands and Survey decided to subdivide the remaining land into 35 residential sections and sell them for holiday homes. But within a couple of decades erosion on the spit started to escalate again. Large chunks of the foredunes disappeared during the 1968 storm that sunk the *Wahine* off the coast of Te Whanganui-a-Tara Wellington, and further erosion occurred soon after.

Property owners tried to stop the tide with iron and mānuka protection walls, old car bodies and pieces of concrete. But these efforts ultimately proved futile. In April 1976, the first house was claimed by the sea. More than half of the 35 sections ended up being fully or partially underwater. A 1976 publication by the Ministry of Works and Development concluded that 'future development of any kind on Ōhiwa Spit should be discouraged as the consequences will obviously be disastrous'.

Government pitched in to assist the land-owners. After all, it was at least partly to blame as Lands and Survey had subdivided the area in the first place. Around 10 of the land titles were converted to public ownership in exchange for land at Matatā, where some of the houses were relocated. Other sections at Ōhiwa remained in private ownership, mainly because the owners were not willing to accept the government's offer.

During the 1980s, the erosional cycle reversed, and the Ōhiwa spit started building back up. The earlier sections gradually re-emerged out of the sea. Accretion on the spit was helped by the combined efforts of the Ohiwa Beach Coast Care Group and the Whakatōhea Māori Trust Board, who had planted the dunes with native sand-binding vegetation.

By the early 2000s, the price of coastal property had skyrocketed, especially beachfront land. This prompted the Brownes, who owned 2 Ohiwa Beach Road, to put their section on the market. The *Property Press* advertisement waxed lyrical about the merits of the property, stating,

> words can't describe the unspoilt beauty that surrounds this beach front section. Be in awe of the sunsets reflected in their magnificence off the Pacific Ocean. Walk the kilometres of pristine golden sands, often sharing the beauty only with your shadow. This stunning stretch of East Coast remains relatively undiscovered. Giant Pohutukawa trees watch over the beach line, revealing their glory when in flower to these select few privileged enough to be part of this unique environment. Walk from this property directly onto shimmering sands and a safe swimming beach ... this could be your own slice of Pacific Paradise ... Let the Pacific waves sing you an evening lull-a-by to end a perfect day. Don't delay. This section was last for sale 35 years ago.

When EDS first heard about plans to build a house on land that had recently emerged from the sea on the Ōhiwa spit, the team headed to Ōpotiki to investigate the proposal. Shown here is EDS Chief Executive Gary Taylor meeting with local residents on the spit. RAEWYN PEART

The advertisement gave no hint as to why the property had not changed hands for 35 years — it had been submerged! And this would not be the only time such submergence was likely to occur. Scientific investigations had revealed that the rebuilding of the spit was likely only temporary and it was predicted to soon reverse. Instead of the waves singing a gentle lullaby, as described in the advertisement, they would likely create a nightmare when they got too close again.

But that didn't appear to deter the new owner, Ohiwa No. 2 Limited, which was planning to build a holiday home up on poles. Its application to the council was accompanied by a coastal hazard risk assessment which indicated that the building would likely need to be removed within just 13 years due to coastal inundation risk. A worst-case scenario put this at just over six years. The application was heard by the Ōpotiki District Council in August 2006. EDS mounted strong opposition to the proposal, along with local residents who wanted to see the area retained in its natural state. They were incensed that their work to naturally rebuild the dunes had potentially enabled the development of it.

Despite the opposition, the council granted consent. It had no choice. The house was a controlled activity under the district plan, and therefore consent had to be granted, albeit subject to conditions. As a condition of consent, the council required the applicant to remove the building when erosion reached a distance of 30 metres from the house.

EDS appealed the council's decision to the Environment Court but knew that it was on slippery ground. The society claimed that the applicable rule in the district plan was outside the scope of the RMA, and therefore ultra vires, but EDS's lawyers advised that the argument lacked a strong legal foundation because the legislation was not strong on this point. The chances of winning on appeal were slim. This prompted EDS to seek settlement with the council; if it couldn't stop the current development, it wanted to make sure that no more would follow.

As a result of negotiations, the council agreed to prepare 'Plan Change 1 — Proposed Ohiwa Spit Coastal Hazard Zone', which was notified in 2008. This made development on the spit a non-complying activity (which could be turned down) instead of a controlled one (which couldn't). EDS submitted on the plan change, seeking prohibited rather than non-complying status for buildings in the extreme hazard risk zone, so that an application could not be made in the first place. But the council was unmoved on this point and it also made an exception for small relocatable buildings.

Not satisfied, EDS appealed the decision to the Environment Court, but the parties settled before the matter was heard. EDS agreed to the non-complying activity status (as opposed to prohibited) and the council agreed to remove the exception for relocatable buildings. The overall result provided a high level of protection for a vulnerable area of New Zealand's coastline. The result seems even more important 15 years on, with climate change rapidly increasing coastal hazard risks and talk of managed retreat.

FURTHER NORTH, there were concerning proposals to develop another low-lying sandspit, this time at Ngunguru. The Ngunguru Sandspit lies on the Tutukākā coast east of Whangārei. Before EDS became involved, there had been numerous attempts to develop the 152-hectare sandspit after it was sold out of Māori ownership in 1964. But these attempts had been vigorously opposed by the Ngunguru Sandspit Protection Society, as well as Ngāti Wai, who considered the spit to be wāhi tapu as many warriors had been killed there.

There had also been numerous attempts to bring the sandspit into public ownership, but none had succeeded, largely due to differences in expectations around its value. In 1993, for example, the Department of Conservation looked to buy the sandspit for the government valuation of $518,000 but the owner at the time, Auckland company director Bob Green, claimed to have independently valued the spit at $3 million and his land agent claimed it was worth over $5 million.

Things came to a head when the sandspit was purchased in 2003 by Landco Land Development Limited, a development company part-owned by the country's then wealthiest family, the Todds. A year later the spit was again up for sale but no suitable buyers were found. So Landco proceeded to formulate development plans in order to realise the value of its investment. At a series of community consultation events held in late 2005, Landco revealed that it wanted consent to enable 350 houses to be located on the sandspit.

EDS became actively involved in influencing the future of Ngunguru Sandspit in early 2007 after Landco announced its intention to lodge a resource consent application for the development. The news also prompted the formation of the Ngunguru Sandspit Protection Society. EDS directors visited the site and on 16 January 2007 Gary Taylor wrote in the *New Zealand Herald*,

> The idea that in the 21st century we could consider putting urban, residential density development on a low-lying, erosion-prone sandspit, which has

> outstanding landscape and wildlife values, is bizarre. Sea level rise and increasingly violent storms from climate change ought to be enough to knock it on the head without having to go into an assessment of the broader environmental effects. A response from the developers has been to avoid the lowest land and to offer to put dwellings on poles.

Landco eventually agreed to swap its land on the Ngunguru Sandspit for the former Napier hospital site and the bulk of the spit was secured in public ownership. However, the landward end of the spit and adjacent mountain (at the top of the image) is still in private ownership, with its future uncertain. GARETH COOKE/ KAHURANGI PROJECT

The article made it clear that EDS would staunchly oppose the development, alongside the local community group. It concluded by urging Landco to 'act responsibly and abandon the project'.

Fortunately, that's what Landco ultimately did. In June 2008, Todd Land Holdings

Limited acquired the balance of Landco and at that point offered to swap the land on the Sandspit for developable land elsewhere. It is not clear what motivated this decision, but the high level of local opposition, backed by an organisation such as EDS which was willing to fight the development all the way through the courts, likely played some part. It took some time to find a suitable site, but on 25 August 2011, the government announced that the sandspit had been acquired by the Department of Conservation and that the Todd Property Group, in a related transaction, had purchased the former Napier hospital site. The Ngunguru Sandspit proper was finally protected for future generations to enjoy and appreciate.

However that wasn't the end of the story. The land swap didn't include the landward end of the spit or the adjacent bush-clad mountain, both of which remained in Todd's ownership. This was eventually sold to development company Templeton Commercial Limited.

About a decade after the main sandspit was secured into public ownership the Ngunguru Sandspit Protection Society negotiated a conditional sale and purchase agreement with Templeton for the remaining land at a price of $3.6 million. The society then set about raising funds to complete the purchase, but eventually had to pull out of the deal when it failed to secure sufficient money. Local hapū have also expressed their wish to obtain ownership of land due to its cultural importance and presence of sensitive wāhi tapu sites. This is still a work in progress.

WELL AWARE OF THE importance of getting the planning framework right, soon after the society's re-establishment, EDS became heavily involved in the development of the Far North District Plan. The plan covered some 1700 kilometres of coastline, including Bay of Islands and five major Northland harbours.

The Far North District Council first released a proposed plan in October 1996. This provoked strong opposition from the rural sector, particularly over efforts to identify and protect significant natural areas (such as remnant native bush) on private land as this would restrict farming activities such as clearing indigenous vegetation. Nearly 4000 submissions were lodged, most in opposition. There were calls to dump the plan and sack the council. The plan was ultimately withdrawn and a second proposed plan released in late 1998.

It was not until July 2003 that the council issued decisions on submissions to the plan and these were still too lax when it came to coastal subdivision. EDS appealed the decisions to the Environment Court, and the matter was first heard in February 2005, with an initial focus on coastal subdivision rules. Former EDS director Ian Cowper acted for both EDS and Bay of Islands Coastal Watchdog in seeking more rigorous controls on coastal development. Opposing them was businessman Craig Heatley (through Bentzen Farms Limited), who was supporting the council in its wish to provide a more flexible and less restrictive approach. Eventually the parties reached a settlement which, although not going as far as the environmental groups had sought, significantly strengthened the subdivision controls in the rural coastal zone.

The first case to be decided under the new provisions was at Ngaiotonga, a stretch of rugged, undeveloped coastline south of Cape Brett. Landscape architect Stephen Brown described the area as 'one of New Zealand's

In the early 2000s EDS helped strengthen the subdivision rules for the Northland coast. These were first tested in relation to a development proposal at Ngaiotonga, a particularly distinctive and striking part of the coast. Shown here is the coastline just to the north of Ngaiotonga, including Elliot Bay. RAEWYN PEART

most distinctive, striking and instantly appealing coastlines'. The areas around the proposed development site were 'particularly important as a truly majestic, wild and scenic coastal landscape'.

Recognising such values, the area was identified as an outstanding natural landscape in the district plan. Notwithstanding this, it was here that Kingsford Barker and Associates Limited, a company owned by North Americans Jeffrey and Debra Anlauf, proposed to create a rural-residential subdivision providing 14 house sites.

The last Māori landowner of the 57-hectare property had been Wiremu Pita. Pita had farmed the land with others, but in 1973 he sold it, ostensibly due to concerns over council rates. His family still retained ownership of land immediately to the south. A subsequent owner of the Ngaiotonga property had bulldozed large parts of it, destroying a pā site and damaging much of the bush. It was then sold to the Anlaufs.

The new owners took a much more sensitive approach. They hired experienced designers and engaged with the local community, providing employment as part of restoration works on the land. The local marae became supportive of the proposal. But when the Pita whānau became aware of what was sought for their ancestral land, they were alarmed and mobilised as kaitiaki to protect the area's important cultural values. EDS became involved to support the Pita whānau case. The society was concerned that the development would despoil a stunning part of the coast.

At the Environment Court hearing, Wiremu Pita's grandchildren gave extensive evidence about the attachment the whānau had with the land, through building homes and burying whenua there, roaming over it, collecting rongoā and flax and gathering kaimoana. M. C. Pita 'spoke of the strong cultural bonds that he and other family members felt for the land [and] … gave details of why he thought those cultural bonds would be offended by the proposed subdivision'. In particular, houses would be seen from important fishing places around the bay.

The applicant had brought in a bevy of consultants to support the proposal, including award-winning architect Pip Cheshire, planner Brian Putt and landscape architect Denis Scott. To assist in persuading the court of the merits of the proposal, the Anlaufs had commissioned the development of a complex visual model, where aerial photographs and other images such as maps and plans were overlaid on a three-dimensional representation of the area. This was designed to depict the impacts of the development.

Putt argued that the development should be approved because the buildings would be carefully sited with mitigating vegetative planting, and there would be controls on building heights, colours and materials. Pip Cheshire described the proposed establishment of a design review board and application of design guidelines to the buildings. In addition, the developer proposed to covenant and protect over 90 per cent of the property in bush and to undertake replanting and weed and pest management. Pedestrian access to the coast, through the site, was proposed for tangata whenua but (on their request) not for the public.

Despite all these mitigation proposals, buildings would still be visible from coastal waters and they would also be seen on the skyline and close to headlands. Stephen Brown, who gave landscape evidence for the Pita whānau, thought the proposal would result in an excessive concentration of development on an exceptionally sensitive part of the coastline. It was coastline that was 'as close to pristine as one could find outside a national or regional park'. He did not give much credence to the applicant's commitment to replant the area as a form of offset, as the site was naturally regenerating in any event. Gary Taylor and Shay Schlaepfer presented submissions on behalf of EDS, supported by planner Peter Reaburn.

In the end, the Environment Court was unswayed by the sophisticated analysis produced by the applicant. Due to the toughening up of the district plan provisions, through the earlier efforts of EDS and others, the proposal had become a non-complying activity. This meant it had to pass through a 'gateway' test before approval could be legally granted. The applicant needed to show that either the adverse effects of the development on the environment were minor, or that it would not be contrary to the objectives and policies of the plan.

It was clear that the adverse effects of the proposal would be more than minor, due to its visual and cultural impacts, and with the toughening up of the council plan's objectives and policies, it was not compliant with those either. Despite being sensitively designed, this type of development was simply not appropriate on such a wild and culturally significant part of the coastline.

Ultimately the court declined consent. However, in issuing an interim decision, it left open an opportunity for the applicant to come back with a modified proposal with fewer lots. This didn't happen. It later emerged that the Anlaufs did not yet have the required consent from the Overseas Investment Office to purchase the property.

As a result of the case, the natural and cultural values of the ancestral land of the Pita whānau remained protected. In addition, the case set an important precedent for the protection of the coast; proposals to undertake planting of native bush were not sufficient to justify urban development on pristine parts of the coastline.

WHILST SEEKING TO head off inappropriate coastal development on the ground, EDS was also busy urging government to strengthen national policy so that decisions on coastal subdivisions were not left to the whim of small and often impecunious councils. Then Conservation Minister Chris Carter had commissioned a review of the New Zealand Coastal Policy Statement in 2003. It was undertaken by Massey University planning academic Dr Jo Rosier, who delivered her findings in May the following year. She concluded that the policy statement had only been partially effective in influencing the content of district plans and subsequent land-use planning decisions in the coastal environment. Her report recommended that the document be formally reviewed.

It took another four years, but finally the Conservation Minister (now Steve Chadwick) released a new proposed Coastal Policy Statement in 2008, and appointed an independent board of inquiry to consider it. The board was chaired by Environment Court Judge Shonagh Kenderdine. In November that year, a new National-led government came to power, bringing a more pro-development approach to government policy. Notwithstanding the change in government, the board of inquiry process continued.

In the end, 539 written submissions were lodged on the draft Coastal Policy Statement, indicative of the level of public interest in coastal matters. The board of inquiry held 30 days of public hearings. EDS was given a full day to present its detailed submissions, with Raewyn Peart presenting the results of her coastal research undertaken on behalf of the society.

Peart's work comprised the first focused investigation into coastal development in recent years and so was influential in the process. She highlighted the enormous pressure in many parts of the country for coastal development, the inability of many councils to manage it, and the poor outcomes that were the result.

Eventually, in July 2009, the board of inquiry delivered its report and recommendations. They were received by yet another new Minister of Conservation, National's Tim Groser. The board of inquiry proposed a much more directive and protective coastal policy than the 1994 version. However, this did not necessarily align with the new government's more pro-market direction. The recommendations disappeared into the bowels of government. Late in 2009, when the board of inquiry's report had still not publicly emerged despite EDS requesting it under the Official Information Act, EDS complained to the Office of the Ombudsman.

In early 2010, another Minister of Conservation, Kate Wilkinson (the fourth minister in little over two and a half years) and the Ombudsman had still not made a

EDS had a strong influence on strengthening the New Zealand Coastal Policy statement to better protect the country's coastline (including places such as Ocean Beach, Hastings shown here) from urban development.
RAEWYN PEART

decision on EDS's complaint. In March 2010, the Ombudsman indicated a desire to hear Minister Wilkinson's views before ruling on whether the report should be released or not. Months passed and the report was still not forthcoming. So, on 6 May 2010, a member of the board of inquiry who had been a former Minister of Conservation in the Fourth Labour Government, Philip Woollaston, decided to publish the report. He did so because he believed it had been buried by government due to it being seen as anti-development.

In its report, the board of inquiry held that several of EDS's requests were outside its terms of reference, including placing a moratorium on subdivision under a certain size and undertaking a national exercise to identify outstanding natural landscapes. However, it did recommend strengthening policies in other key areas, in line with what EDS had sought, and this represented a major step forward in coastal protection.

When the government finally promulgated a new Coastal Policy Statement in October 2010, the board of inquiry's recommendations had been watered down in some important respects. More emphasis was put on providing for development within the coastal environment, with specific new policies inserted for ports and aquaculture. But some provisions emerged unscathed. These included the requirement to avoid adverse effects on outstanding natural landscapes and areas of significant vegetation and habitat. Councils were required to identify areas which were 'inappropriate' for development. In addition, a new category of 'outstanding' natural character was introduced for protection.

The new version of the New Zealand Coastal Policy Statement came into force in December 2010. It was eventually tested by EDS in the highest court in the land in the King Salmon case (see chapter 14), and turned out to be much more protective than people initially thought.

Peter Reaburn, a planner who was closely involved in EDS's efforts to strengthen the policy, recalls,

> I was very happy to be involved in EDS's advocacy in relation to the review of the coastal policy statement. We went along and set up quite a substantial, well-considered and well-based submission on what the statement should have in it.
>
> EDS had two major planks. The first was, don't produce a policy statement that's woolly. Have a statement that gives appropriate direction so when councils give effect to it they know what they are giving effect to. Second, the government should go further and actually look at its own role in mapping and classifying important parts of the coast. The result was success in terms of the first plank but not the second one. We all know the benefits that it has achieved since the Coastal Policy Statement 2010 came into force. I think EDS can be quite proud of its involvement in that exercise.

Landscape architect Di Lucas, who also gave evidence to support EDS's case, remembers the society being very well received by the board of inquiry. She considers the New Zealand Coastal Policy Statement to have been one of the better policy documents produced by government. 'For EDS I helped ensure that we got ONLs [outstanding natural landscapes] in the marine as well as terrestrial areas and that natural character was addressed across the spectrum. We also got a raft of attributes recognised, so it was not just the view that was highlighted as important, but biophysical associative and perceptual attributes.'

The year before the new Coastal Policy Statement was finalised, EDS published a book authored by Raewyn Peart titled *Castles in the Sand: What's Happening to the New Zealand Coast?* in partnership with Craig Potton Publishing. As described in the April 2009 edition of *EDS News*, the book described 'what the coast means to New Zealanders and how New Zealanders are losing those things about the coast they love the most. It also talks about putting things right, about better caring for the coast so that future generations can continue to cherish and enjoy New Zealand's superb coastal heritage.'

To assist with the research and images for the book, EDS borrowed a helicopter and pilot to fly the Northland coast, and notable New Zealand landscape photographer and publisher Craig Potton agreed to come along and take a series of images which EDS could use at no charge. Several of the images are reproduced in this book.

Bay of Islands developer Peter Cooper hosted the book launch at his Britomart waterfront development. Former Planning Tribunal Judge Arnold Turner attended, as did former Hauraki Gulf Maritime Park board member Jim Holdaway. Auckland Mayor John Banks was the guest speaker and he spoke eloquently about

In order to gain a better appreciation of development pressures on the Northland coast, and to document them, EDS borrowed a helicopter to fly the area. Landscape photographer and publisher Craig Potton (second from right) joined the trip to take photographs. Coastal planner Greg Hill is far left and EDS Chief Executive Gary Taylor second from left looking into the camera. RAEWYN PEART

his love of the Auckland coastline. The book was well received, with reviewer Graham Adams in the Air New Zealand inflight magazine *KiaOra* writing, 'this excellent book is a lament for what the author sees as widespread despoliation of our seaside ... It's a comprehensive and highly readable analysis of the problems that have developed as the number of coastal dwellings has soared, with councils often incapable or unwilling to control their proliferation or the damage they do.'

—ooo—

ONE OF EDS'S LONGEST-running campaigns to protect sensitive parts of the country's coastline has been focused on New Chums Beach (Wainuiototo Bay). EDS became involved in the protection of the beach soon after the society's re-establishment in 1999, and the effort continues more than 20 years later.

New Chums Beach is located on the east coast of Coromandel Peninsula, just north of the small beach settlement of Whangapoua. There is no public road to the beach and access requires wading across the Whangapoua estuary, scrambling over rocks along the shore-line of the southern headland, and then crossing privately owned land to reach the beach.

Despite the difficult access, the beach is very popular and highly regarded. It has been named one of the most beautiful beaches in the world by *Lonely Planet* and *National Geographic*. It features a native bush-clad

In 2009 EDS published (in association with Craig Potton Publishing) the book *Castles in the Sand*, which chronicled the importance of the coast to New Zealanders and the development pressures on it. Developer Peter Cooper hosted the book launch at his Britomart waterfront development. Shown here is EDS policy director Raewyn Peart at the launch with former Planning Tribunal Judge Arnold Turner (left) and former Hauraki Gulf Maritime Park board member Jim Holdaway. EDS

escarpment fringing an attractive, curving, white-sand beach.

Once the last bastion for local iwi Ngāti Huarere, the land has both historical and cultural significance. It features Māori taro gardens, burial sites, gardening areas and a fishing hole. The beach was called 'New Chums' after the men who travelled from Auckland to work on the kauri log booms in Wainuiototo Bay. As new arrivals, they were called 'new chums'. The name stuck and was subsequently used for the bay they were working in.

The Denize family bought land within the New Chums catchment and along the northern end of Whangapoua in 1944. It was known as Te Pungapunga Station. The family operated a sheep and beef farm for some years, but eventually the increasing value of coastal property during the 1990s undermined the economics of farming there. The high land value also created difficulties with the inter-generational transfer of the farm. Although the Crown looked at buying the property during the 1980s for a coastal park, the purchase did not go ahead. So, in 1999, the Denize family sold the farm to Auckland businessman Clive Currie.

Currie was not a farmer and he was keen to develop the land. He applied to the Thames-Coromandel District Council for consent to undertake a residential subdivision on a ridge overlooking New Chums Beach and a second subdivision overlooking the northern end of Whangapoua. Both were strongly opposed by the Whangapoua Beach Ratepayers Association and EDS. Ngāti Huarere also opposed any development of the land due to the historical significance of the area.

The ridge application was rejected by council but, due to lax provisions in the district plan, consent had to be granted for the five house lots overlooking Whangapoua. There were fears that this was just the first stage of a more intensive subdivision. Currie subsequently gained consent for a house to be built next to New Chums Beach. Gary Taylor expressed concern about Currie's plans, and in particular the scale and changing

New Chums Beach with its native-bush-clad escarpment and curved white-sand beach has been named one of the most beautiful beaches in the world. However, the beach and land behind it is in private ownership and has been subject to myriad development proposals. RAEWYN PEART

nature of what was proposed. He was worried that the development was undercapitalised and could end up being 'cheap and nasty'.

Currie's plans never came to fruition and he eventually looked to sell. EDS and the ratepayers association tried to persuade the government to purchase the property, but to no avail; they had recently made the Waikawau Bay purchase (described earlier) and indicated that enough was enough in terms of land acquisition on Coromandel Peninsula.

So, in 2002, Currie sold the land to Queenstown landscape architect and developer John Darby, his business associate and friend Christchurch-based funds manager George Kerr, and Wellington-based investment banker Ross Mear and his wife Deidre. The land was in five main certificates of title, with two additional small pā site and access lots. The purchasers were formally Northern Coastal Land Trust Holdings Limited (under the control of Darby) and Galt Nominees Limited (under the control of Kerr). The purchase price was rumoured to be around $4.5 million. The multiple land ownership arrangements made future dealings with the property complex.

In 2006, Kerr sought consent for a substantial house site just behind New Chums

With the land behind New Chums Beach largely unprotected under the Thames-Coromandel District Plan, EDS worked closely with the owners of the land to look for a solution that would protect the beach and its backdrop while enabling the owners to recoup their investment. RAEWYN PEART

Beach, which the council was legally obligated to grant. In 2008, Darby obtained consent for a five-lot subdivision on land facing Whangapoua, where the original homestead and farm buildings were located. But this did not affect New Chums Beach itself.

Concerned to head off the construction of the Kerr house next to the beach, Thames-Coromandel Mayor Philippa Barriball contacted him direct and offered the use of council planners and architects to work on a compromise development. In exchange, Kerr surrendered the consent for the beach house.

The exercise resulted in a development plan comprising 20 house lots, none of which were claimed to be visible from New Chums Beach, although some would be visible from the Whangapoua settlement. They were to supersede the five house lots already consented by Currie. The development plan also included putting 291 hectares into public ownership, which would include a walking track around a significant stand of kauri.

EDS had worked closely with the developers in formulating the scheme. Gary Taylor had found John Darby and his associates to be well-intentioned people who were trying to reach a good outcome. The proposal was seen as the best possible result given that the government had not wanted to acquire the land and that the weak planning framework under the RMA seemed unable to protect the beach whilst it was in private ownership.

The plans were publicised in 2009 and were

not well received by the public. In particular, the ratepayers association did not want any development to be visible from Whangapoua. Many people wanted to see the land behind New Chums Beach protected from any development at all.

The proposals prompted the formation of 'Preserve New Chums for Everyone' as an incorporated society with the goal of preserving the beach in a pristine state for future generations. The founder was Linda Cholmondeley Smith, whose family had formerly owned much of the land around New Chums Beach and still owned the southern headland which was protected by a Queen Elizabeth II National Trust covenant. The public outcry prompted the owners to look for a different solution.

EDS continued to work with Darby and Kerr to come up with something that would enable them to recoup their investment in the land while protecting the beach for the public. What emerged in 2010 was a proposed private-public partnership, whereby the government would be offered the opportunity to buy the New Chums Beach catchment and adjoining area of kauri forest in order to create a public reserve. However, soon after the proposition was made, the first Christchurch earthquake hit. Realising that the government would now be short of money due to the high recovery costs of the earthquake, the offer was withdrawn before it could be formally declined.

Darby and Kerr then fell back on the original 20-lot proposal. This now included a large, covenanted protection lot and an esplanade reserve to give the public legal access to New Chums Beach for the first time. Once publicly notified, the proposal attracted some 1000 submissions. In response to the high level of opposition, the landowners presented a revised application for a 12-lot subdivision, but this, too, was later withdrawn.

Then, in 2013, an application for a 3-lot subdivision on the 61-hectare beach title was lodged. As a controlled activity under the district plan, council was legally obligated to grant consent. Fortunately, before this happened, the application was put on hold.

When the proposed Thames-Coromandel District Plan was publicly notified in December 2013, the rules that applied to development of land behind New Chums Beach were up for deliberation. It took some years to get through the council submission and hearings process, but eventually the rules ended up before the Environment Court with a hearing held in mid-2019.

The main focus was on the details of a site-specific structure plan that was to be applied to the property to strictly control the number and location of house sites, and to protect the balance of the property where no development could occur. There was disagreement about the number and location of house sites, and how much protection and bush planting should be required along with any residential development.

Then, out of the blue in early November 2021, Gary Taylor heard that the 30-hectare land title over the northern headland of New Chums Beach had been put up for mortgagee sale by receivers acting on behalf of the Bank of New Zealand. Fortunately, none of the house sites proposed in the structure plan were located on the headland. It also had no legal access other than below high tide along the beach. Both these factors potentially reduced its value, making it more affordable.

When he heard about the sale, Taylor swung into action. At last there was an opportunity

The northern headland of New Chums Beach was purchased by the New Zealand Coastal Trust, an entity established by EDS for the purpose of coastal protection. Shown here viewing the site after the purchase are, from left: Trust Chair The Hon Peter Salmon KC; Celia Caughey, who has a holiday home at Whangapoua and helped raise money for the purchase; EDS Chief Executive Gary Taylor, who initiated the bid; and part-owner of the remaining privately owned land in the New Chums catchment, Ross Mear. RAEWYN PEART

to bring at least part of the land behind New Chums Beach into public ownership. The tender closed on 23 November so there were only a few short weeks to raise the necessary finance.

As Taylor later observed,

> it was an ambitious task. I started by making public comments designed to hose down the value of the property and deter well-healed offshore buyers who were rumoured to be interested. This incurred the wrath of the Bank of New Zealand. But slowly its key people came around to seeing that public ownership would be the best outcome. Some very well-motivated individuals came to the party and offered to underwrite the purchase, pending institutional funders contributing. It was an amazing upwelling of support and showed how people still value the coast.

Taylor worked with Ngāti Huarere, Preserve New Chums for Everyone and the Whangapoua Beach Ratepayers Association to raise enough funds, with the help of the underwriters, to make a realistic tender. Preserve New Chums set up a Givealittle page for public donations. Celia Caughey, who had a holiday home at Whangapoua, made direct approaches to residents in the settlement. In addition, applications were made to the Department of Conservation's Nature Heritage Fund and Environment Waikato's Natural Heritage Fund.

The tender was made in the name of the New Zealand Coastal Trust, an independent

New Zealand Coastal Trust trustees visiting New Chums Beach after the purchase. From left the Trust's Chair The Hon Peter Salmon KC, Gary Taylor and Raewyn Peart.
DAVID HARTMAN

entity which had some years previously been established by EDS as a mechanism to protect coastal land through direct purchase and covenanting. The trust, chaired by retired High Court Judge Peter Salmon KC, and with other trustees Gary Taylor, Raewyn Peart, former Bell Gully law partner David McGregor and Anglican Bishop Richard Randerson, was brought out of hibernation. Jan Chen, a senior associate at Bell Gully, provided pro bono legal assistance for the tender.

The trust lodged a tender for $1.75 million by the deadline of 23 November 2021. This was followed by a second tender bid of $2.15 million, which was ultimately successful. It was hard to believe, but part of the land behind New Chums Beach, and a very sensitive part at that, had finally been secured for the public benefit. It had not been government that had purchased the land, but the public themselves.

The bid had been underwritten by supporters pending sufficient funds being secured. This was achieved when the Margaret Turnbull Estate and John Turnbull Estate jointly contributed 45 per cent of the final purchase price. The Coastal Trust intends that the land be managed forever in the public interest and it will be protected by a covenant from the Queen Elizabeth II National Trust.

Just before the tender for the headland closed, the Environment Court finally released its interim decision on the broader property. This approved, in principle, a structure plan for the land which would provide for 25 house sites. Much of the land, which is outside the area designated for house sites and which includes the beach escarpment, is proposed to be identified as conservation area. Appeals against the Environment Court decision have since been filed in the High Court.

As this book goes to print the future of the balance of the land at New Chums Beach has still not been settled. Never giving up, EDS has set its sights on acquiring the escarpment land behind the beach.

11. Natural landscapes under threat

RURAL DEVELOPMENT PRESSURES were not just on the coast but had begun spreading out into many of New Zealand's other iconic landscapes. EDS was keen to protect what were some of the country's last undeveloped areas before it was too late. Landscape protection was the topic of EDS's first national conference and its first policy report. It was also the impetus for litigation in areas as diverse as Taupō, Kaipara, Heretaunga Hastings and Te Manahuna Mackenzie Country.

AS WELL AS BECOMING involved in a wide range of cases to protect individual landscapes under threat from development, EDS looked at ways to raise the profile of landscape loss more generally. One idea was to convene a conference. As Raewyn Peart recalls,

> we toyed with the idea of holding a conference to highlight landscape protection issues. However the financial risk seemed too high. EDS had virtually no money and would have to guarantee payment for venue hire and catering as well as for a conference organiser. We decided to apply to the New Zealand Law Foundation to bring an overseas speaker out to the conference, as this was one of the few relevant categories of funding available. Eventually the Foundation came back and said 'yes', so we thought we had better run a conference so the overseas speaker has something to attend.

Organising a big conference with virtually no funds was a particularly stressful enterprise and there were more than a few sleepless nights. EDS needed to raise sponsorship for a novel event for which there was no track record.

After much effort, Gary Taylor managed to cajole a group of sponsors to come onboard, and he also persuaded the New Zealand Institute of Landscape Architects to partner with EDS to convene the conference. The New Zealand Landscape Conference finally went ahead on 25 and 26 July 2003 in the Bruce Mason Centre at Takapuna, Tāmaki Makaurau Auckland. Malcolm Grant, a New Zealander who at the time was Professor of Land Economy and Pro-Vice-Chancellor at the University of Cambridge, flew in, with Law Foundation funding, to address the conference.

The conference was opened by Minister of Conservation Chris Carter. He began his presentation stating:

> As we all know, New Zealand has some of the most remarkable landscapes in the world. Our mountains, our beaches, our grasslands and our forests shape and influence our culture, our beliefs and our vision as a society. What we do to them speaks volumes about who we are as a nation.
>
> For this reason, I think it is important that conferences like yours periodically take place to assess the state of the landscape and our efforts to control the way we alter it. This is particularly so because humans are good at recognising and responding to radical change but less attuned to gradual change, and it is my view that gradual change can often have the biggest impact on the landscape.

Te Warena Taua, an elder of Te Kawerau ā Maki, provided an overview of Māori perspectives on landscape and emphasised the deep interwoven nature of Māori society with the natural world.

Sculptor John Edgar spoke of a 'landscape with too few lovers' reflecting on New Zealand artist Colin McCahon's *Northland Panels*, which inscribed that phrase and provided a sad indictment of the way New Zealanders had treated the land.

Previous pages: Strong development pressures on iconic rural areas such as the Wakatipu Basin shown here, prompted EDS to focus its efforts on landscape protection. This was the topic of its first New Zealand landscape conference and first policy report. RAEWYN PEART

The Waitākere Ranges, which were threatened by strong urban development pressures, was one of the places discussed at the first EDS landscape conference and the ranges became one of the five case studies that EDS researched for its first landscape report. CRAIG POTTON

Sessions of the conference then focused on landscape and culture, how well landscapes were being managed, international experience and possible solutions for New Zealand.

The conference was well attended and garnered a great deal of interest. In his summing up of the two days, landscape architect Simon Smale called it a 'highly energised conference' and one that had been 'prompted by a growing sense of general unease, of loss and regret, engendered by the nature, rate and scale of some of the changes that are occurring in coastal and high country landscapes, but also in places like the Waitākere Ranges'. He also noted that the conference had been a 'very Pākehā affair' and he highlighted the potential 'that is still largely latent in the Māori world view — for the advancement of our national understanding of landscape, which is as much about relationship, association, meaning and perception as it is about landscape's physical reality'.

In the final plenary session, a series of resolutions were passed by acclamation of those present, including that the conference recorded its concerns over the potential outcomes of the high-country tenure review and affirmed that New Zealand's outstanding and significant landscapes need better protection and management.

EDS had achieved its main aim of raising awareness of the threats to the country's landscapes and the need to better respond. In the aftermath of the conference, landscape provisions were strengthened in many district plans. As EDS director and law professor Barry Barton observed,

> I think EDS scored an enormous success in turning landscape around from something that was dismissed as too hard, or too subjective, into a situation where all of a sudden everyone was saying 'we know how to do landscape and we're working hard on it in our district plan review'. It moved quite quickly from impossible to possible. EDS mainstreamed it. I think we can claim that.

The conference was the first of what was to become a regular event. Since then, EDS has convened a national environmental policy conference most years. The conferences have developed into a high-profile national environmental summit, drawing together a broad spectrum of interests, including politicians, officials, resource management professionals, scientists, Māori and community groups.

AS WELL AS HIGHLIGHTING the issue of landscape degradation through the conference, EDS was keen to undertake an investigation into how councils were managing or mismanaging the country's important places. From what EDS had observed, councils were really struggling in the face of strong development pressures. EDS wanted to understand why this was the case and what could be done to remedy the situation. Gary Taylor looked to obtain resourcing for the project.

It turned out that raising funding for a little-known organisation that lacked a current track record was not easy. Government departments and councils were reticent to give money to an environmental group that might criticise their performance or even take them to court. But eventually EDS managed to cobble together small amounts from the Ministry for the Environment, Department of Conservation and Ministry for Culture and Heritage. In addition, the Thames-Coromandel and Whangārei district councils chipped in with financial support for local case studies. It was enough for EDS to get started. The rest would rely on pro bono effort. Former lawyer Raewyn Peart agreed to head the project.

Peart sought to investigate what was happening on the ground in some of the coastal and lakeside development hotspots; the Waitākere Ranges, Te Tara-o-te-Ika-a-Māui Coromandel Peninsula and Whangārei District in the North Island and the Wakatipu Basin and Te Pātaka-o-Rākaihautū Banks Peninsula in the South Island. She went out into the field to talk to people from many different sides of the landscape controversy, including councillors, residents, environmental activists and landowners. Peart wanted to find out what was happening on the ground.

The project went well until Peart arrived on Banks Peninsula. She had arranged to meet up with representatives of Federated Farmers to find out more about the landscape protection controversy that had been raging there. Some years previously, the North Canterbury Branch of Federated Farmers had threatened to take the Banks Peninsula District Council to the High Court if it progressed the landscape protection provisions in its proposed district

Above: Whangārei District was one of the case-study areas for EDS's first landscape report. But by the time EDS started to investigate what was happening in the district, much of the coastline had been despoiled by poorly managed urban sprawl and coastal development on ridgelines, as shown here at Tutukākā. CRAIG POTTON

Below: Coromandel Peninsula was another case study where coastal development had been rampant. Shown here are two large houses built very close to the beach at Hahei, which detract from the public's ability to enjoy the natural coastal environment. RAEWYN PEART

plan. In response, the council paused the plan-making process and established a rural task force to consider the matter. Farmers were given a strong voice on this task force, so it was not surprising on reporting back to council in 1999, that it recommended a considerable watering down of the landscape protection proposals, which the council largely accepted.

Above and right: Banks Peninsula, including Akaroa Harbour shown above right, was a case study for EDS's first landscape report, though EDS policy director Raewyn Peart did not get a positive reception when she first visited the area to interview local farmers. The farming sector had persuaded the Banks Peninsula District Council to retreat on its landscape protection initiatives and did not welcome the involvement of an environmental organisation from Auckland. RAEWYN PEART

Having achieved this coup, the rural sector was not necessarily pleased to see EDS, which was known to advocate for stronger landscape protection, arrive in town. Peart got a frosty reception. As she recalls, 'I remember meeting these two ferocious women from the farming sector on Banks Peninsula. They both really ripped into me, questioning what I was doing there. I was pretty naïve at the time and not used to that level of animosity. I went pale, mumbled something and quickly left the room.'

Peart and Taylor also headed overseas to investigate alternative models of landscape protection, including those in the United

Kingdom, Canada and the United States. Those countries had developed various mechanisms to put an extra protective layer over special landscapes which were wholly or partly in private ownership. EDS had concluded that the RMA was not strong enough to protect important landscapes on private land, so it was keen to promote a 'third way', whereby further protection was provided.

The research culminated in a report titled *A Place to Stand: The Protection of New Zealand's Natural and Cultural Landscapes*, which was launched in May 2004. A press release accompanying the launch explained that it was 'the first comprehensive look at the management of New Zealand's natural and cultural landscapes under the Resource Management Act.' The press release went on to state:

> The situation is far worse than we originally thought ... There is poor understanding of what landscapes we should protect and why. There is no national picture, a very uneven regional picture and a confused local picture. This means there's uncertainty for the environment, for businesses and developers, for councils, and for the public. The research shows a number of breakdowns in how we identify and protect important landscapes. We

> have no government policy on what landscapes we should be protecting and for what purpose. Regional councils need to identify regionally significant landscapes and not many are doing that. At the district level, councils need help to deal with these complex landscape issues.
>
> Effective landscape management is also hindered by the widespread use of planning techniques which fail to adequately address cumulative effects across a wide area, an 'approval culture' which means that most resource consent applications are given the go ahead, and a lack of monitoring of the outcomes for important landscapes so that decision-makers have little idea of the impacts of their decisions.

The report proposed that a package of remedies be implemented, including the promotion of better practice, the promulgation of a strong national policy statement on landscape, and new special-purpose legislation to support the piloting of protected landscape models.

EDS hoped the report would raise the profile of the issue as well as motivate politicians to intervene and fix things before it was too late. Although landscape protection did subsequently receive more attention, it would still be a long battle to improve outcomes for the country's important landscapes. And it was a battle largely fought at the council level and in the courts.

THE PUSH BACK that had occurred on Banks Peninsula in response to the council inserting landscape protection measures into its district plan was also experienced in other rural districts. A notable case was Kaipara District, which included a large portion of the land bordering the Kaipara Harbour, the largest harbour in the southern hemisphere. It also included a stunning stretch of north-east coast at Mangawhai.

The Kaipara District Council had embarked on a full review of its district plan in 2005. This had included commissioning a landscape study from Littoralis Landscape Architecture, which had identified 23 outstanding natural landscapes within the district.

The process of developing plan provisions then proceeded slowly. It was not until 2008 that council officers began to consult directly with landowners and the reaction was not what they had hoped for. Farmers were outraged that the council was seeking to control what they did on their land and mobilised to put a stop to it.

The farmers were successful, and at a council meeting on 26 November 2008 the council resolved to take 'no further action on the identification of, and consultation on, Outstanding Landscapes in the Review of the Kaipara District Plan'. The reasons recorded for the recommendation were that:

> Council needs to concentrate on key strategic elements of the District Plan including the promotion of economic growth and environmental wellbeing. Council considers the resources of managing the outstanding landscape process outweighs the benefits to the rate payer and the community. Council notes the issue is likely to be dealt with in the public and Environment Court processes of the District Plan Review.

The Kaipara District includes a large portion of the land bordering the Kaipara Harbour, including at Paparoa shown here, but the council failed to provide protection for important harbour landscapes. ROB SUISTED/ NATURESPIC

As a result of this resolution, no further work was undertaken on landscape matters by council planners, and landscape material was excluded from the proposed district plan which was notified in 2009. The table of contents for the proposed plan included a Chapter 18, which had notionally been to address landscape matters, but on turning to that chapter the reader was greeted with the advice that 'this chapter is intentionally blank'. For this reason, the matter became known as the 'blank pages' case. Although councils were known to have notified proposed plans which were deficient in a number of respects, in EDS's experience none had removed an entire chapter before notification.

The RMA required councils to protect outstanding natural landscapes as a matter of national importance. So EDS considered the failure to include any provisions on this topic in the proposed plan to be in contravention of the statute. The council was in effect saying

that people should submit on what should be in the landscape chapter and the Environment Court should then write the chapter, thereby filling in the gap that the council had left. EDS thought this was an abdication of legal responsibility.

Although the Kaipara District has many important landscapes, including those at Mangawhai shown here, the council deleted all landscape provisions from its proposed district plan before notification. Through EDS's efforts, the council eventually notified a landscape variation to its proposed district plan. CRAIG POTTON

What was even more extraordinary was that the council appeared to be getting away with it. No other government agency, and in particular the Ministry for the Environment, which was tasked with environmental oversight, was holding it to account. EDS decided to act. The society hoped to not only put a shot across the bows of Kaipara District Council but provide a warning to other councils that might be tempted to shirk their environmental responsibilities.

Just before Christmas, on 23 December 2009, EDS lodged proceedings in the Environment Court seeking declarations that the council's actions were unlawful. It argued that the council had a duty, under the law, to notify a proposed plan that recognised and provided for outstanding natural landscapes and features. Greg Milner-White from Kensington Swan acted for EDS and Peter Reaburn provided supporting planning evidence.

The following April, while the court proceedings were pending, Kaipara District Council resolved to notify a variation to its proposed plan by the end of September 2010 in order to address landscape management. But there was no certainty that the council would meet this deadline and so EDS pursued its legal challenge.

As it turned out there was no hearing. The Environment Court made its decision on the papers. It was largely an open-and-shut case. Judge Newhook found that 'Not only, as a matter of fact, did the council resolve to down tools on the PDP's [Proposed District Plan's] landscape chapter, it has to be acknowledged that the operative plan is also very incomplete on the topic.' The judge also found that there was no basis under the RMA for arguing that EDS or other parties should do the council's work for it and create a landscape chapter through the submission process.

The court issued the declaration as sought by EDS and the council eventually notified a landscape variation to its district plan on 2 December 2010. Outstanding landscapes within the district finally had some level of protection.

But perhaps more importantly the case sent a warning to other councils. They now knew that EDS was watching, and if they failed to deliver on their statutory duties, they could also end up in embarrassing court proceedings. The court decision made it clear that councils could not just decide to avoid difficult matters, they had a statutory duty to protect the environment.

EDS director Professor Barry Barton found the guardian of the process role that EDS played in the case particularly satisfying. He noted, 'That sort of litigation is hugely effective, it sends shock waves around councils.' It was not the only time that EDS had to hold a council to account.

THE MAPARA VALLEY lies to the west of the Taupō CBD. It is a rural pocket extending in a north-easterly direction from Whakaipo Bay on Lake Taupō. The valley was created during the various Taupō eruptions when extrusions of rhyolite rock formed hilly ridges bounding a wide valley floor. Mapara Valley Preservation Society stalwart Sarah Foreman remembers the first time she saw the valley when, with her husband Peter, she was looking to buy a home in Taupō. 'We rounded a bend to go down a steep hill. My heart stopped as the vista opened up of a largely unspoilt valley surrounded by jagged hills and embankments. You could feel its history, a sense of space and the past geological forces of nature at work.'

Sarah and Peter Foreman moved to the rural farming area in 2002. They had not been there long before they heard about proposals to fundamentally change the valley. In particular, a company called Lakeview Ventures was seeking to build a town of some 1331 residential dwellings and associated commercial activities on one of the large valley farms close to the lake. The Foremans were horrified that the magnificent rural landscape they now lived in could be developed in such a way. They made contact with the Mapara Valley Preservation Society and became active members.

Sarah Foreman was worried that their under-resourced group would be severely out-gunned by the development company, which had engaged top lawyers to run its case. She scoured the internet looking for someone to help and chanced upon the EDS website. She rang Gary Taylor and he agreed to send planner Peter Reaburn to take a look. As Foreman recounts,

> Peter was to travel from Auckland and I agreed to meet him in the valley. I had no clue what to expect when his car pulled up. He got out, a tall lanky figure with a wide boyish smile. He exuded professionalism. I instantly felt calm. I took him to the edge of the volcanic rim on Whakaroa Road and he looked out at the magnificent view. He nodded and understood. He 'got it'.

By the time EDS got involved in late 2005, the Taupō District Council was being flooded with subdivision applications. Developers were seeking to meet the strong demand for holiday homes in the area and were taking advantage of a particularly weak district plan. Nutrient limits within the Lake Taupō catchment were also encouraging farmers to look to alternative uses for rural land. As well as fending off individual developments, EDS saw an urgent need to strengthen the rules in the district plan itself.

As Peter Reaburn recalls, 'It was an old traditional plan that said you could basically do anything in the rural area. There was very little control on activities. The subdivision rules were hopeless.' EDS filed an appeal on behalf of the Mapara Valley Preservation Society to strengthen the plan's provisions. This was the start of a long process of plan development, with the council notifying a series of variations to fix the plan.

As Reaburn later related:

> I remember sitting with Shay [Schlaepfer, EDS's in-house lawyer] in a room of landowners, developers and council people. We were pushing things completely on our own. I remember feeling that our argument was a really good one. We pushed and pushed and in the end it made so much sense that we got some respect from around the room. Then the Council changed its position and was on our side.

In the end, the district plan provisions were considerably strengthened. But developers rushed to get rural subdivisions approved under the old rules and EDS was involved in several efforts to stop them.

One of the biggest development proposals in the Mapara Valley was that of Lakeview Ventures, the development Sarah and Peter Foreman had heard about soon after moving to the valley. The property in question occupied much of the lower valley, with part of it being less than a kilometre from Lake Taupō and forming the backdrop to public areas along the lakeshore.

The company originally applied for a 1331-lot development, but then put that on hold and sought consent to subdivide the same property into a much less intensive development of 57 lots, which complied with the district plan's controlled activity rules. It was this smaller subdivision that went to a hearing.

As Peter Reaburn later explained in his evidence for the Mapara Valley Preservation Society, 'the "drivers" of the subdivision are obvious. Rather than a subdivision design that respects the environment, the parent site has simply been divided into 4-hectare blocks in an attempt to meet the controlled activity

RAEWYN PEART

Shay Schlaepfer joined EDS as its first in-house lawyer in 2007, after graduating with a double degree in law and science. At just 22, she was keen to deploy both her law and science skills in a future career and EDS provided a unique opportunity to combine the two.

As well as fielding numerous calls as part of the community advice service, Schlaepfer coordinated the EDS litigation programme. Some cases were identified through the community advice service when matters raised by community members revealed serious public issues. In addition, EDS subscribed to a regular report listing all plan changes and resource consent applications across the country. Schlaepfer would review the reports to identify any issues that EDS should consider becoming involved in. She then worked with external lawyers and experts who assisted EDS to run the cases.

At that time EDS was still a tiny organisation struggling for funding and could not afford a functioning office. As Schlaepfer recalls, 'it was a massive learning curve. EDS was very small and there were no other lawyers in-house. I lived in a small unit in Point Chevalier and set up an office in my lounge. EDS had a room in an old villa in Richmond Road which had a lot of boxes and a printer. I would work there for the day if I needed to print stuff out.'

Schlaepfer was involved in the Taupō cases as well as the Ngaiotonga development. 'I was managing about 30 cases, liaising with lawyers and experts about evidence and hearings. It was fast-paced, fascinating and a fantastic grounding for my legal career. We did lots of site visits and I got to see much of the country. I also saw the political side of things. I felt very grateful for the opportunity.'

Schlaepfer left EDS in late 2009 to further her legal career, first as a barrister with Simon Berry and then as a solicitor at Brookfields. She also spent several years overseas. Schlaepfer, now with young children, returned to EDS initially part-time, contributing to its policy and litigation work, including the Waitākere Ranges landscape case study, the Mataura ouvea premix case (ouvea premix is a by-product from aluminium dross processing), a review of the Land Act and EDS's Wildlife Act reform project. Schlaepfer's role has expanded over time, and she has recently been appointed EDS's chief operating officer. Schlaepfer has been an EDS director since 2009.

thresholds ... Houses will be uniformly spaced at about 200-m intervals.' In Reaburn's view, it was a poorly designed, 'chequerboard' subdivision that would fundamentally change the rural nature of the valley for good.

By applying for the maximum number of lots permissible as a controlled activity (which by law must be consented), the company may have been seeking to ratchet up the 'permitted baseline' of effects, to pave the way for its much larger 1331-lot application, which had not been withdrawn. In EDS's experience it was not unusual for developers to seek to 'game' the rules in order to progressively gain consent for large developments.

As anticipated by the applicant, the council

granted consent to the 57-lot proposal on the basis that all the sections met the minimum 4-hectare size requirement. It appeared that little could be done to stop the development. Notwithstanding this, the Mapara Valley Preservation Society appealed the consent to the Environment Court. EDS acted for the society with Kensington Swan partner Rob Enright providing legal representation assisted by EDS lawyer Shay Schlaepfer. Peter Reaburn gave planning evidence in support.

EDS became involved in stopping urban development in the Mapara Valley in association with the Mapara Valley Preservation Society. It meant engaging in numerous plan change and resource consent proceedings. The society managed to hold the line until the development pressure eased with the 2008 Global Financial Crisis. The valley remains rural. RAEWYN PEART

Schlaepfer has strong memories of the case. There was a lot at stake but the prospect of winning seemed remote. In the run up to the hearing she had contacted Enright and

asked ‘do you need my help with preparing the case?’ He had responded ‘not really, see you at the hearing’. She recalls, ‘he came down the day before the hearing and prepared the submissions that night. The next morning he was really excited. He’d uncovered a flaw in the proponent’s plans.’

What Enright had uncovered when he had looked at the subdivision plan in detail was that three lots which were to be accessways were less than 4 hectares. This technically meant that the proposal did not meet the criteria for a controlled activity when it was lodged with the council. The court agreed and considered it as a discretionary activity, which meant that all the impacts of the proposal would be considered. At that point, the council, which had earlier defended its decision to grant consent, changed tack and supported the Mapara Valley Preservation Society in opposing the development.

The court ultimately upheld the society’s appeal and quashed the 57-lot consent. It was a significant win. But it still left the earlier application for 1331 lots to be resolved. It was now 2009 and the Preservation Society and EDS had been fighting development of the site for three long years. Fortuitously, before the second application could be heard, the company was placed into liquidation. To this day, the large farm has not been subdivided and it is used for cropping and dry-stock farming.

The court decision did not only stop development of that site, but it also had a chilling effect on other applications. In particular, Tukairangi Valley Properties had applied for consent to subdivide a 429-hectare site at the head of the valley into 95 lots but abandoned its plans after the Lakeview Ventures court decision.

EDS became involved in many other zoning and development proposals in the Mapara Valley, seeking to protect the notable rural landscape from urban development. At one stage the council proposed to locate an urban development node in the valley linked via a regional arterial road to State Highway 1. EDS and the Mapara Valley Preservation Society strongly opposed this proposition, including the designation for the arterial road.

The last subdivision case EDS was involved in was a proposal by Sade Developments to develop the peninsula jutting out into Lake Taupō at Whakaroa Point. The company proposed to subdivide a 344-hectare headland farm into 86 residential lots.

This time, in a toughening of its approach, the Taupō District Council turned down the application. Sade Developments appealed to the Environment Court and the Mapara Valley Preservation Society became a party supporting the council's decision. John Burns and Shay Schlaepfer acted for the society. Ultimately the court upheld the council's decision to decline consent, although it did provide an opportunity for the developer to come back with a scaled-down application. This didn't happen.

This was the last significant development application to be considered under the old lax district plan rules. The gold rush had been beaten back, largely as a result of EDS standing behind a group of local residents. As Gary Taylor noted, the court decisions 'taken together, should signal the end to ad hoc speculative rural development in the Taupō District.' In 2008, the Global Financial Crisis hit New Zealand and demand for holiday housing in Taupō evaporated.

EDS had put considerable resource into a multitude of Taupō cases. It was a substantial effort for a small organisation and a severe drain on scarce resources. It was an example of EDS offering to help a small local group and then being drawn into a multitude of related issues in one area. In reflecting on his time working on the Taupō cases, Peter Reaburn observed, 'EDS's arguments are always sound ones. I think over time there has developed a preparedness by councils, in particular, to listen to what EDS is saying. It has got better and better to the point now that if EDS is involved, you listen. Back then, it was still a learning exercise for some councils.'

Shay Schlaepfer fondly remembers her involvement in the Taupō cases.

> Back in those days EDS didn't often pay for hotels, so I stayed with Sarah Foreman and her husband Peter or with other members of the Preservation Society. Living in their homes gave me an appreciation for the depth of feeling that these people had towards the land, and how greatly their lives were being affected by the proposed developments. I was not only their lawyer, but I became a trusted friend.

Sarah Foreman recounts,

> the EDS team would congregate at our house after visiting the site, gobbling down soup, warming up and debriefing. The Aucklanders found Taupō freezing (which it was!). Shay would come down and stay and I struggled to find interesting recipes for her vegetarian palate usually ending up with boring roast veges. She would tease me about my lack of imagination with my cooking. It was lovely to have her as we could chat informally about planning and environmental subjects. I have wonderful memories of the unity for a cause.

The Mapara Valley remains largely rural and undeveloped.

—ooo—

Te Mata Peak is one of Hawke's Bay's most iconic landscapes and it is of great cultural significance to Ngāti Kahungunu. Depsite this, the council consented the construction of a walking track on its eastern face (shown on the right of the image) which created a prominent zigzag scar. RAEWYN PEART

IN OCTOBER 2017, Craggy Range Winery started to build a 2.4-kilometre walking track on the eastern face of Te Mata-o-Rongokako Te Mata Peak, one of Te Matau-a-Māui Hawke's Bay's most iconic landscapes. Once the track started to become visible, there was an outcry from iwi and the public about the prominent 'scar' that the construction works had created on the flanks of the peak. The wide zigzag track was a very jarring footprint on the notable natural feature. Highlighting the extent of the visual intrusion, *The Dominion Post* headline in December read, 'It looks like Te Mata Peak has had open heart surgery.'

It subsequently emerged that Hastings District Council had granted consent to the track without public notification. This was on the basis that its effects would be 'no more than minor'.

The decision had been made by a council official without the knowledge of the mayor or the councillors. Mana whenua had also not been

RAEWYN PEART

Rob Enright qualified with a Bachelor of Laws (Hons) from the University of Auckland in 1995. He joined Kensington Swan in 1997 and became a partner there in 2001. In 2010, Enright left to form law firm Kirkland and Enright, where he was a partner. In 2013 he became a partner at DLA Piper (previously DLA Phillips Fox), before becoming a barrister sole in 2014.

Enright was a director of EDS for a period in the early 2000s. He has acted for EDS in numerous cases, including the Lakeview Ventures development in Taupō, Te Mata Peak, Plan Change 13 and Simons Pass Station in the Mackenzie Basin, the Akaroa marine reserve case, the Ruataniwha dam case, the Mataura ouvea premix proceedings, the King Salmon case and the Man O'War case, and he managed to devise innovative legal arguments that won many cases for the society.

informed about the proposal despite the deep cultural sensitivity of the site: Te Mata Peak was known as the final resting place of Rongokako, the founding ancestor of Ngāti Kahungunu.

The peak had been identified in the district plan as an outstanding natural feature and landscape with a rating that made it one of the most significance landscapes in the district. But despite its cultural significance to iwi, it had not been identified in the plan as a wāhi taonga. When he heard about the construction works, Ngāti Kahungunu chair Ngahiwi Tomoana harshly criticised the track.

After EDS lawyers looked into the matter, they concluded that the grant of consent on a non-notified basis was unlawful. In December 2017, the society wrote to the council and Craggy Range Winery saying so, and threatening judicial review. In response, the winery wrote back to EDS agreeing to remove the track. But then things started to unravel. Craggy Range subsequently claimed it could not remove the track without further consent and indicated that it would prefer that the track remain. Some members of the public were very opposed to the track, but a countervailing campaign had been started by those who wanted to see it retained.

EDS decided it was time to take legal action. It worked alongside the Waimarama Marae to file proceedings in the High Court in July 2018, which argued that the Hastings District Council's decision to approve the track was unlawful, and sought an order for its removal.

EDS asserted that the council had made two errors of law in approving the application. First, the application should have been publicly notified because of its national importance as an outstanding natural landscape and feature of cultural significance and, secondly, the council had failed to apply the correct legal tests under the RMA when granting approval. Rob Enright

COURTESY OF PETER REABURN

Peter Reaburn qualified with a Bachelor of Resource and Environmental Planning from Massey University in 1980. He then worked as a planner for the Upper Hutt City Council, Waitematā City Council and Waitākere City Council. In 2000 he went into private practice as a planner with Cato Bolam Associates, where he currently still works. He is a keen waka ama paddler.

Reaburn became involved in assisting EDS through a friendship he developed with Gary Taylor when Reaburn was planning manager at Waitematā City Council and Taylor was a councillor.

Reaburn has been a 'go to' planner for EDS for many years. He has assisted the society with numerous cases including Karikari Peninsula, Ngaiotonga, Kaipara District, Mapara Valley, Mackenzie Basin, Ruataniwha Dam, Te Mata Peak and Marlborough Sounds King Salmon. He also gave evidence to support EDS's submissions to the board of inquiry on the New Zealand Coastal Policy Statement and co-authored the EDS guide *Strengthening Second Generation Regional Policy Statements*. He has recently assisted EDS with its policy work on the reform of the conservation management planning system.

and EDS in-house lawyer Madeleine Wright acted for EDS on the matter.

The lodgement of proceedings brought the council to the table and in October 2018 it announced that it would restore the top part of the track, which was deemed unsafe, under emergency provisions of the RMA with the remediation to be undertaken and financed by the council. It then also sought consent from itself to remediate the lower parts. A year later, in October 2019, work started to fill in the track and restore Te Mata Peak to its former glory. It may take some years for the scars in the landscape to disappear.

But things did not end there. The case served to highlight the importance of protecting Te Mata Peak and in October 2022 the council adopted a plan change that now prohibits new buildings on its eastern face. This will help protect the significance and value of this outstanding natural landscape for both current and future generations.

The battle to protect New Zealand's significant natural and cultural landscapes continues. Once these landscapes are developed, they are effectively lost forever.

12. Protecting the Mackenzie

THE MACKENZIE COUNTRY is the popular name for an expansive area of high-altitude land located in the central South Island that is surrounded by mountains to create a geological basin. Carved out by ancient glaciers, the basin features three natural lakes — Ōhau, Pūkaki and Tekapō. It is the only place in New Zealand where the entire intact glacial sequence can be seen, from glaciers high up in the Southern Alps through to large areas of glacial moraines, and extensive outwash terraces and plains.

THE MACKENZIE IS subject to climatic extremes, being very dry in summer and very cold in winter. It is this challenging climate, coupled with the varied glacial topography, that has enabled a diverse and distinctive biota to evolve. In particular, there are communities of rare desert-adapted indigenous plants on the basin floor. Originally called Te Manahuna, the area used to be largely covered in scrub and indigenous grasslands. It was a rich source of kai for Māori over many centuries.

Things changed after a contested land transaction, known as Kemp's Deed, enabled the Crown to wrest the area from Ngāi Tahu. Much of the land was then leased to runholders. More than a century of pastoral farming, which involved regular burning of the tussock and oversowing with introduced grasses, resulted in the loss of many indigenous plant communities. However, under low-intensity merino sheep and beef production, the Mackenzie Basin had largely retained its expansive brown aesthetic. That was until irrigation started to 'green' the area.

The drivers of irrigation stemmed back some decades. From the early 1950s up until the 1980s, New Zealand's largest network of hydroelectric power stations was constructed in the Mackenzie Basin.

The works had a significant impact on farming practices, as much productive flat land on the basin floor was flooded when lake levels were raised and new hydro lakes and canals established. Runholders received little financial compensation for the loss of their pastoral lease land but, in order to offset some of the impacts, the government agreed to provide access to irrigation water in order to enable farm intensification on the remaining flats. This water started to be taken up during the 2000s, and the face of the valley floor gradually turned from brown to green as introduced irrigated pastures proliferated.

At the same time as demand for irrigation increased, there was a push to freehold low-lying Crown pastoral leasehold land in the Mackenzie Basin through a process called tenure review. This sought to transfer sensitive and erodible high-elevation land into full government ownership (to become public conservation land) in return for the more productive lower areas on the basin floor being freeholded, enabling more intensive farming and other economic uses.

Overlooked in the design of the tenure review policy was that it was these low-lying areas that the basin's distinctive desert-loving indigenous flora and fauna needed to survive. This brought farming in the Mackenzie Country directly into conflict with biodiversity conservation as well as landscape imperatives.

EDS'S INVOLVEMENT in the Mackenzie was prompted by a 2009 proposal to house 17,850 dairy cows on Ōhau Downs Station. The project was based around the concept of 'cubicle farming', whereby the cows would be housed indoors for nine months of the year in 20 wintering sheds and be milked by robots. Effluent from the sheds was to be collected into large effluent ponds holding up to 77 million litres. Up to 1.7 million litres of effluent was to be discharged to pasture daily, equivalent to the volume held by more than 110 milk tankers.

When he first heard about the proposal, EDS

Previous pages: Lake Tekapō is known for its stunning turquoise colour caused by glacial silt washing down the Godley and Macauley rivers. The Mackenzie Basin is the only place in New Zealand where the entire intact glacial sequence can be seen. RAEWYN PEART

When the EDS team first heard about the cubicle farming proposal they travelled to the Mackenzie Basin to investigate the situation first-hand. Shown here are EDS team members meeting on-site with cubicle farm proponent Richard Peacocke. From left, EDS chief executive Gary Taylor, EDS events director Fiona Driver (partly hidden behind Taylor), then Richard Peacocke and on the far right is EDS in-house lawyer Kelsey Serjeant. RAEWYN PEART

Cubicle farming involves housing cows indoors for much of the year and collecting their effluent, which is then sprayed back onto the fields. Shown here is a cubicle farm barn housing dairy cows in Canterbury. The volume of effluent that would be produced from the Ōhau Downs Station proposal was estimated to be equivalent to a city of 250,000 people. RAEWYN PEART

Chief Executive Gary Taylor was alarmed at the effects of the very large barn structures on the values that made the Mackenzie Country such an iconic landscape. He was also concerned about the impacts of large-scale dairying in such an inhospitable climate. In his view, the desert-like and ecologically sensitive environment of the Mackenzie Country was no place for dairy farming, which required large quantities of water and discharged copious amounts of effluent. His view was confirmed by landscape architect Di Lucas who helped EDS review the project.

By the time EDS became aware of the cubicle farming proposal, the Waitaki District Council had already granted consent on a non-notified basis for the buildings and the conversion of indigenous tussock grassland to irrigated rye grass. This was despite the cubicle barns being around 30 metres wide, 6 to 7 metres high and up to 150 metres long. Such large structures, in EDS's view, would significantly intrude on the open, vast and uncluttered landscape.

In March 2010, EDS filed proceedings in the Timaru High Court judicially reviewing Waitaki District Council's decision to grant the resource consents. Natasha Garvan, who was a junior lawyer seconded to EDS from Bell Gully at the time, was involved in the case. She had a close family connection as her father was a councillor on the Waitaki District Council. Garvan recalls, 'Rob Enright had picked up on the fact that the council person who made the decisions didn't have the right delegations. So the applicant conceded and the judicial review never went ahead. That felt like quite an effective outcome for EDS.'

The part of the Mackenzie Basin under the management of the Waitaki District Council, including Lake Ōhau shown here, is highly vulnerable to farming intensification as this is still a permitted activity in the district plan.
RAEWYN PEART

While the judicial review proceedings were progressing, Environment Minister Nick Smith used his ministerial powers to 'call in' an application to discharge effluent from the cow sheds, which had been lodged separately. This was on the grounds that water quality issues in the fragile Mackenzie Basin were of national importance. A call in meant that the matter would not be heard by the council in the first instance, but by a ministerially appointed board of inquiry or the Environment Court.

As the minister explained in a written statement, 'the effluent from these intensive farms is equivalent to a city of 250,000 people and raises quite legitimate questions over the long-term impacts on the water quality in the Mackenzie Basin'. At that time, the population of the largest settlement in the basin, the town of Twizel, was little over 1000.

Once the applications were publicly notified, 4852 public submissions were received, indicating the significant level of public concern about the proposals. Many of the issues raised related to animal welfare, but there was also concern about impacts on the Mackenzie Country's stunning landscapes. Once the matter had been called in, the cubicle farming proponent withdrew the effluent discharge application.

Former Environment Minister Nick Smith recalled,

> there was broad public unease in the community about the scale of dairy expansion in the fragile area of the Mackenzie Country. It was the EDS legal challenge over both the land and water allocation consents that moved the dial from the point of view of sending a shot legally through the court action, and also politically to the government that a reset was required. That resulted in me making a decision to call in the application. The applicant then recognised it wasn't going to fly and walked away.

The proposal looked to have been stopped in its tracks, although in October of the same year EDS heard that the cubicle farm proponents had applied for new planning consents. EDS threatened more High Court challenges if the applications proceeded. This seemed to be enough to finally put the proposal to bed.

As a result of this initial and successful foray into Mackenzie Basin development issues, EDS became alerted to the very weak planning framework that applied to the area, and the need to broker a better solution for the future of the basin.

The area was currently being managed by two small and under-resourced entities — the Mackenzie and Waitaki district councils. As Gary Taylor said to media at the time, 'clearly there has been a failure of public policy at all levels. The government has failed to provide national guidance ... and the two district councils have failed to develop coherent and effective district plans.'

In order to help shift things onto a more positive track, EDS decided to convene a one-day symposium in Twizel. The aim was to bring stakeholders together to consider the future of the Mackenzie Country and find a common way forward. Wanaka resident Bruce Jefferies, who at the time was Vice Chair of the New Zealand World Commission on Protected Areas, was very supportive of the idea. EDS collaborated with his organisation, as well as Forest and Bird and the International Union for the Conservation of Nature, to convene a one-and-a-half day symposium on 26 and 27 November 2010.

The symposium itself was held in the Twizel

Events Centre on Saturday 27 November, with a field trip around the basin undertaken the previous afternoon. Earlier in that week, Federated Farmers announced that it would boycott the event. Vice-president Donald Aubrey called it an 'Imposium'. Mackenzie Federated Farmers branch chairman John Murray objected to people from Tāmaki Makaurau Auckland and other areas coming down to impose their views on farmers. He was reported in the *Otago Daily Times* as stating 'Why should they have the right to come down here and tell us what to do?'.

Gary Taylor responded through the media, 'I don't see what can be achieved by boycotting a conversation. It's very immature, and I thought we had got past that.' This initial sparring captured the imagination of local media, which continued to profile the various competing views throughout the week.

It was therefore with some trepidation that the small EDS team travelled to Twizel to convene the symposium. The group consisted of EDS chief executive Gary Taylor, policy analyst Raewyn Peart, in-house lawyer Kelsey Serjeant and event organiser Fiona Driver. They arrived on Thursday night and booked into the Mackenzie Country Hotel. Little did they know that it was the very night, and the very hotel, that the local branch of Federated Farmers held its regular meetings.

After checking in, the EDS team met in a corner of the hotel bar for a drink. Across the other side of the room were the Federated Farmers meeting attendees, who were also relaxing over a beer. Included in their group was cubicle farmer proponent Richard Peacocke.

Peart can clearly remember the tension in the room as the two groups sought to ignore each other. The jousting between them through the media just days before was fresh in everyone's minds. Eventually, Richard Peacocke walked across the room and offered to buy Taylor a drink. The tension was broken and an uneasy truce followed.

Earlier in the week, around 120 people had registered for the symposium, but as a result of the high-profile media coverage, around 220 actually pitched up, and almost half were locals. This far exceeded EDS's expectations, and overwhelmed the capacity of the Twizel theatre, with people spilling out into the aisles. Although national and local Federated Farmer representatives failed to attend, many farming families in the region made low-profile appearances. Also in attendance were some key decision-makers: Environment Minister Nick Smith, the mayors of Waitaki (Alex Familton) and Mackenzie (Claire Barlow) and local National MP Jacqui Dean.

By the end of the symposium there was a coming together of the different parties, and broad agreement that local people needed to take the dialogue forward. The two district mayors agreed to progress things. Nick Smith indicated that government money might be forthcoming to support a collaborative process. Overall, despite the tense beginnings, the symposium had been a real success.

The collaborative process did eventually go ahead. Jacqui Dean brokered an agreement to set up the Mackenzie Sustainable Futures Trust, which funded and oversaw the collaborative deliberations of the Upper Waitaki Shared Vision Forum. Participants included representatives from Federated Farmers, Dairy New Zealand, Merino New Zealand, the Mackenzie Irrigation Company and other irrigators, tourism operators, salmon farmers, community boards and residents associations. Fish and Game and a range of environmental organisations, including the Mackenzie

The Mackenzie symposium convened by EDS in Twizel in November 2010 was highly successful and brought a wide range of sectors together to chart a common vision for the Mackenzie Basin. Shown here are attendees gathered during the field trip held the day before the symposium proper. RAEWYN PEART

Guardians (a local community group), Forest and Bird and EDS were also members of the forum, which first met in February 2011.

In-house lawyer Kelsey Serjeant represented EDS during the early months of the collaborative process. She recalls flying down to Christchurch every couple of months, being picked up by Jen Miller from Forest and Bird and driving down to Twizel. They would stay for a couple of nights and have a multi-day meeting with the other stakeholders. 'For me it really did highlight some of the difficulties of policy-making, being in the same room as people whose livelihoods and interests were very different.'

After three years of meetings, forum members reached an accord called the 'Mackenzie Agreement' which was launched in Twizel in May 2013. The agreement proposed that 100,000 hectares of the Mackenzie Basin be protected as a drylands park for landscape and biodiversity conservation and that 26,000 hectares be intensified. It also proposed establishing the Mackenzie Country Trust which would negotiate joint management agreements with landowners and high-country run holders.

Such agreements would involve payments in exchange for managing the land in a way that was sympathetic to biodiversity and tussock grassland protection objectives. All participants had agreed to a common way forward. It

seemed to be a turning point for the basin. Nick Smith:

> to the credit of EDS, having drawn a line in the sand through court action, it then sought to engage and bring together the parties to work on a better alternative plan. It would have been easy for EDS, largely perceived as an Auckland-based NGO, to alienate the high-country farming community in a place like the Mackenzie. It showed real political smarts that following court intervention, EDS not only participated, but actually drove the desire for a collaborative approach that actually got tourism, farming, hydro, ecological and recreational interests actively discussing a better future for that particularly stunning and fragile part of New Zealand.

Although the Upper Waitaki Shared Vision Forum managed to reach agreement on a future for the Mackenzie Basin, the government declined to fund implementation and so it languished. Meanwhile the basin's outstanding landscapes (including land near the Ben Ohau Range shown here) were coming under increased pressure from irrigation and farming intensification. RAEWYN PEART

But the agreement's implementation required significant government funding and this was not forthcoming. The environment portfolio was now in the hands of Minister Amy Adams, who did not have the same enthusiasm for collaborative processes as the former incumbent, Nick Smith. The Mackenzie Country Trust was not established until 2016. Meanwhile intensification of the Mackenzie Basin grasslands proceeded apace.

While the Upper Waitaki Shared Vision Forum was deliberating, changes to the

EDS was in the courts for years fighting for stronger protection of the Mackenzie Basin's iconic landscapes, including the desert-like Grays Hills shown here.
RAEWYN PEART

Mackenzie District Plan which applied to the northern and less-developed portion of the Mackenzie Country were proceeding through the Environment Court.

Plan Change 13 had been publicly notified by the Mackenzie District Council in December 2009, at the same time as the cubicle farm controversy was erupting. Prior to the plan change, there were no effective planning controls over pastoral intensification within the Mackenzie Basin, it being classified as a permitted activity (meaning no consent was required).

When the council first notified Plan Change 13 there had been significant opposition from landowners. This prompted the council to appoint independent commissioners to hear submissions. The commissioners recommended a significant watering down of the proposals, which the council accepted. This spurred the beginning of a long series of Environment and High Court appeals that lasted almost a decade, as the future of the Mackenzie Country was hotly contested. The Environment Court alone issued 11 separate decisions on the matter.

EDS was not initially involved in the Plan Change 13 hearings. However, when the matter first reached the Environment Court, the judicial panel issued an interim decision

expressing concern at the lack of ecological evidence presented to it by the parties. To help address the gap in information, the court explicitly asked the registrar to send a copy of its interim decision to EDS (along with the Director-General of Conservation and Commissioner of Crown Lands) so that the society could consider whether it wanted to become a party. This was highly unusual and was, in effect, an invitation from the court for EDS to become involved in the proceedings. It underscored the high regard with which EDS was now seen.

In response, EDS applied to become a party to the proceedings in July 2012. The application was strongly opposed by Federated Farmers and others who were keen to keep EDS out of the court case. But not surprisingly, given the earlier overt invitation from the court, the application was eventually approved.

EDS then spent five long years battling out the future of the Mackenzie Basin in the courts, until Plan Change 13 was eventually finalised in December 2017. There were numerous mediations and hearings. The EDS case was led by barrister Rob Enright and EDS in-house lawyer Madeleine Wright. They worked alongside other concerned groups including Forest and Bird and the Mackenzie Guardians. It was an enormous effort, but the Mackenzie Basin was something worth fighting for, and it paid off in the end.

Landscape architect Di Lucas, who supported the Mackenzie Guardians through the long process, thought it was 'fantastic that all the basin floor is an ONL [outstanding natural landscape] at a district scale. All the different NGOs worked on that. It was hard, it was so hard.'

Dr Susan Walker, a botanist and ecologist from Landcare Research who also gave expert evidence in support of the Mackenzie Guardians' case, later observed 'EDS picked the issue right, other people didn't. Plan Change 13 was going to be the linchpin on which everything turned. It was such a long-running saga and EDS was there as the key appellant.'

Plan Change 13 was not the only focus of EDS's legal team. While the new plan provisions were taking a painfully slow process to finalisation, there remained a loophole in the district plan rules that runholders were exploiting to gain consent to intensify their operations. Large areas of rare plant communities were being irretrievably lost in the gold rush to get consent for indigenous vegetation removal before Plan Change 13 took effect. EDS sought to close this gap.

EDS worked closely with other groups, including the Mackenzie Guardians, to seek stronger landscape protection for the Mackenzie Basin. Shown here is a joint field trip to the basin by the two organisations. From left: Fiona Driver, Di Lucas, Marie Doole, Madeleine Wright, Jen Miller, Gary Taylor, Anne Steven and Bruce Jeffries. RAEWYN PEART

As reported in the March 2016 edition of *EDS News*:

> A recent tour by EDS personnel of the Mackenzie Country was extremely upsetting. Since the completion of the Shared Vision process that charted an agreed way forward amongst stakeholders, dairy conversions and tussock grassland destruction have proceeded at pace. Planning loopholes have been exploited and the council's response has been slow and ineffectual. The values that ignited the nation's concerns when cubicle farming was proposed and elicited thousands of submissions in opposition to landscape loss are being relentlessly eroded.

In November 2016, the same year as the Mackenzie Country Trust was finally launched, EDS filed proceedings in the Environment Court against the Mackenzie District Council seeking an urgent hearing. The society argued that the council was issuing certificates of compliance for vegetation clearance and pasture intensification that were causing significant harm to nationally and internationally significant ecosystems and biodiversity in the Mackenzie Basin. The certificates were being issued without any assessment of the indigenous vegetation affected or how any biodiversity loss might be offset.

In support of the proceedings, Gary Taylor swore affidavit evidence that 'the pace of change had reached a tipping point and extraordinary intervention was necessary to prevent permanent loss' which would render future more protective provisions redundant. Dr Susan Walker gave specialist evidence to support EDS's case. She highlighted that most of the indigenous vegetation that had been cleared since 2014 was of ecological significance to the region.

However, the legal basis for forcing the council to fill the loophole was flimsy at best, and EDS sought to reach an agreement before the matter was considered by the court. The tactic succeeded and, just three days before the urgent court hearing, which was scheduled for mid-December 2016, EDS and the council reached an agreement on how the vegetation loss might be slowed. The council would urgently promulgate a plan change to delete exceptions to the vegetation clearance rules (Plan Change 17). Walker recalls, 'I had the feeling that Maddie [Madeleine Wright] and Rob [Enright] were sweating a bit as they felt it was quite a tough argument. I can remember an email from Rob. It was one of those two-line emails that reeks of relief and emotion. He said "Susan very pleased to say we have settled." It was a lovely moment.'

The parties requested that the court give the

Madeleine Wright completed a Bachelor of Laws and Bachelor of Arts in Political Science from the University of Otago in 2011. She then worked for Brookfields as a lawyer for 17 months before joining EDS as a senior solicitor in January 2015. Wright represented EDS on the Land and Water Forum and the Biodiversity Collaborative Group. She was programme manager for EDS's Climate Change and Business Conference for over four years. She ran EDS's litigation programme and was involved in many EDS cases, including Plan Change 13 in the Mackenzie Basin, the Man O'War Station case on Waiheke Island, Te Mata Peak, and the Horizons One Plan litigation. She also assisted with EDS's evaluation of the outcomes from the RMA.

Wright recalls that joining the EDS team 'was a baptism by fire. In week two I was presenting to the independent hearings panel. From a personal point of view, working for EDS was like having another family that really cares about the same things you do, and is interested in the same things. The level of influence and connection you can have as a younger person through the job is phenomenal. To be able to go to a personal dinner with a government minister and provide advice on issues that are going to make a difference was a great privilege.'

Wright left EDS in December 2018 and worked for the Department of Conservation briefly before joining Berry Simons as a senior associate. In February 2021 she joined Sally Gepp to work as a barrister.

COURTSEY OF MADELEINE WRIGHT

rules in Plan Change 17 immediate effect, which it did. This held the line until Plan Change 13 came into effect in April 2017. Once that happened, further pastoral intensification in the Mackenzie District part of the basin was effectively halted, except for one potential gaping loophole — Simons Pass Station.

When reflecting back on the cases around the plan changes for the Mackenzie Basin, Madeleine Wright observed,

> there were a lot of cases on Mackenzie Plan Change 13 and also different declaration proceedings. What I felt really proud about was working with Dr Susan Walker and getting the commissioners to understand the importance of the ecological aspects of the landscape; that a thumb-sized bit of leaf had an important ecological role that didn't exist anywhere else. This was quite a change in how people thought about the landscape, engaged with it, and appreciated its values. That for me was a very big win.

—OOO—

THERE WAS ONE development in the Mackenzie Basin that particularly raised alarm bells, with even many local farmers considering that it had gone too far. An Ōtepoti Dunedin accountant Murray Valentine and his wife

Barbara (through their company Simons Pass Station Limited), had obtained ownership of the pastoral lease of Simons Pass Station at the southern end of Lake Pūkaki and joint ownership of the freehold of the adjoining Simons Hill Station.

The Valentines were seeking to develop the joint 9700-hectare property into a large-scale dairy operation accommodating up to 15,000 dairy cows. This would make it of a similar scale to the failed cubicle farm proposal, and the largest dairy farm in Australasia. And it was being established in one of most fragile environments in the country.

The development needed around 80 separate consents, including discretionary consents from the Commissioner of Crown Lands, water permits from Environment Canterbury and land use consents and certificates of compliance from the Mackenzie District Council.

The Valentines sequentially applied for the consents over a period of 13 years, which meant that the full scale of what was proposed was not evident to any of the consenting authorities or the public. The Mackenzie District Council alone received 53 applications for resource consents from Simons Pass Station. Many were granted due to the district plan's lenient rules prior to Plan Change 13 coming into effect.

By the time EDS become aware of what was proposed at Simons Pass, it appeared that all the consents had already been granted and it was too late to take any legal action to stop the development in its tracks. Already in hand were consents that permitted the erection of enormous pivot irrigators, oversowing and topdressing, vegetation clearance and the construction of fences, farm buildings, tracks and houses.

On Simons Hill Station large rotating irrigators had already been installed and large dairy 'nodes' with milking sheds and other large infrastructure associated with operating an intensive dairy farm, were under construction. That the development, which was so obviously incompatible with the Mackenzie Country's landscape and ecological values, had got so far demonstrated a serious breakdown in the statutory management framework.

But after EDS carefully examined the detail of the myriad consents that had been granted, it discovered that Simons Pass Station, which was still a sheep and beef farm under Crown pastoral lease, did not have all the consents it needed to enable a full conversion to dairy. In particular, it still required consent from the district council for irrigation and direct drilling of seed to create pasture. By this time, Plan Change 13 had become operative, so Simons Pass Station would potentially need to apply under the new and more rigorous rules.

In fairness to those who already obtained water permits for irrigation, Plan Change 13 included some transitional provisions. They provided that activities for pastoral intensification or agricultural conversion were a controlled activity if a water permit had been granted by Environment Canterbury prior to 14 November 2015. If not, the application became discretionary, with the full range of environmental effects considered and the council able to decline consent.

By 14 November 2015, the regional council Environment Canterbury had approved a water permit for Simons Pass Station. So on the face of it, the consent for irrigation and pastoral intensification was a controlled activity, and the Mackenzie District Council was legally obligated to grant it (albeit subject to conditions).

But the matter did not end there. The regional council's decision had been appealed to the Environment Court, and the water

Simons Pass Station in the Mackenzie Basin was to be developed into the largest dairy farm in Australasia, with infrastructure to provide for 15,000 dairy cows. Seen in this image are the long pipelines being laid across the station for the massive irrigation network and green patches of grass from the pivot irrigation and oversowing of exotic grass species that had already occurred. RAEWYN PEART

permit was not ultimately confirmed by the court until October 2016, well after the transitional period had expired. This raised a legal issue as to whether the water permit had been 'granted' when approved by the regional council or whether it was at the later date after determination by the Environment Court.

There was an enormous amount at stake. If Simons Pass Station got the consent, the dairy conversion could go ahead. But if the consent was treated as discretionary, it was almost certain that consent would be declined, due to the impacts of dairy conversion on the high landscape and ecological values of the area. With the support of EDS, the Mackenzie District Council sought an Environment Court ruling on the correct application of the rule and ultimately the Environment Court found in favour of the council.

There was too much at stake to let the matter rest there and Simons Pass Station appealed to the High Court. EDS successfully sought leave to join the appeal. This time, the EDS legal team consisted of barrister Rob Enright supported by new in-house lawyer Cordelia Woodhouse. Woodhouse later observed, 'it was a bit of a stretch in terms of the legal argument, but we had to do it. There was no other way to stop the

Although Simons Pass Station Limited appeared to have all the consents it needed for the massive dairy conversion in the Mackenzie Basin to go ahead, including building multiple dairy nodes such as the one shown here, a forensic examination of the paperwork by EDS lawyers found that several consents were missing. This enabled EDS to challenge part of the proposed development. RAEWYN PEART

development.' The future of the property turned on the technical point as to whether the word 'granted' in the Mackenzie District Plan meant both 'granted and commenced' under the RMA. Despite the tenuous nature of the EDS case, Enright argued it forcefully, and he ended up persuading the High Court not to overturn the Environment Court's decision. It was a lesson in how cases with wide implications could be won on narrow legal points.

The impact of this decision was highly significant. It meant that the 2500 hectares of tussock land which had been slated to be covered in large pivot irrigators and grazed by thousands of cows, now needed consent and this would almost certainly not be forthcoming. The landscape values of the brown, open grassland area had been effectively protected. As Gary Taylor stated when the court decision was announced, this meant the end of dairy conversions in the Mackenzie Country, at least in the Mackenzie District portion of it.

However, this still left the issue of tenure review of Simons Pass Station and the potential freeholding of the land. A preliminary proposal which had been agreed by Land Information New Zealand (LINZ) and the Valentines had been publicly notified in April 2017. It proposed

When the Simons Pass Station pastoral lease went into tenure review, only a small portion of the land was to be returned to Crown ownership. After submissions by EDS and others the amount of land returned increased from 23 to 56 per cent of the total. Seen here are the glacial outwash plains on the station. RAEWYN PEART

to freehold the bulk of the land (some 4310 hectares or 77 per cent of the total) with a small area close to Lake Pūkaki to be restored to full Crown ownership.

EDS lodged a lengthy submission arguing that the proposal was fundamentally flawed, being based on an incorrect application of the law and inadequate and outdated information. EDS also sought to negotiate directly with the Valentines in order to reach agreement on what tenure review for the property should look like. This was partly successful.

In August 2020, a final proposal was released by LINZ which had been accepted by the Valentines. It was much improved. An additional 1900 hectares would be transferred to the Crown, increasing the proportion of the property being put into full public ownership from 23 to 56 per cent. In addition, a further 90 hectares of freehold land was to be protected by conservation covenants.

The protected areas include an impressive range of glacial landforms including moraine, a meltwater channel, and outwash plains with ephemeral tarns, kettleholes and rock outcrops. These in turn support vulnerable plant species including a tiny forget-me-not and the largest known population of the threatened New Zealand mousetail, an endemic annual plant. The protected area also includes what is known as 'the Necklace', an area of very visible moraine and outwash sequences, as well as alluvial terraces adjacent to the Pūkaki River.

Although development of the property for intensive farming had not been entirely stopped, the area affected had been considerably reduced and important parts of the landscape are now in public ownership.

—ooo—

TENURE REVIEW had taken its toll on the Mackenzie Country as freeholded land was subject to farming intensification. Land Information Minister Eugenie Sage announced in February 2019 that the government had decided to end tenure review. However, properties that were currently in the process would be allowed to continue. One of these properties was located within the iconic Lindis Pass, which provides the gateway to the Mackenzie Country from the south. The pass itself is an outstanding national landscape with distinctive brown tussock-covered hills soaring high above the main highway. Although part of the Lindis Pass is a scenic reserve much of it is farmed under pastoral lease.

One of the leasehold stations bordering State Highway 8, which runs through the Lindis Pass, is Dunstan Downs Station. The station extends from the highest point in the pass down to the junction with the Ahuriri River within the Mackenzie Basin proper. It therefore provides a major part of the 'Lindis Pass experience' for people driving into the Mackenzie Country. It is an area popular with tourists.

In September 2020, LINZ publicly notified a preliminary proposal for tenure review of Dunstan Downs Station. Although under the proposal a large proportion of the station was to go into Crown ownership, the most visually sensitive part, the land fronting the state highway, was to be freeholded. This would open it up to pastoral intensification, plantation forestry and other development, especially as the Waitaki District Plan which applied to the area was particularly weak.

EDS was appalled by the proposal and swung into action. In-house lawyer Cordelia Woodhouse put together a comprehensive submission backed by a statement of evidence by senior landscape architect Di Lucas. EDS sought to have the entire lease brought into Crown ownership.

When the final decision was publicised in April 2022 it was a major win. Virtually all of the station (99 per cent) was to be returned to conservation land, one of the highest percentages of a lease under any tenure review process. One of the best examples of tussock landscapes in the Canterbury region was now protected. All the efforts to get the station into Crown ownership had paid off: it was fortunate that LINZ and the Commissioner of Crown Lands were listening.

On 12 May 2022 the Crown Pastoral Land Reform Bill passed its third reading. The new Act ended tenure review and provided stronger protection for the natural values of the high country still managed under pastoral lease.

EDS litigation in the Mackenzie Country is ongoing. The latest round is to do with Variation 18, which seeks to protect indigenous vegetation in the Mackenzie District part of the basin. EDS is also still keeping pressure on the Waitaki District Council to toughen up its plan provisions.

THE AUDACIOUS PLAN to establish large-scale irrigated dairy farming on the dryland floor of the Mackenzie Country at Simons Pass Station prompted EDS to refocus more broadly on landscape protection issues. During the mid-2010s, threats to New Zealand's landscapes other than urban development (which had been the major issue during the 2000s) had emerged. It was not only pastoral intensification and irrigation that was threatening fragile high-county landscapes but also wildling pines.

In addition, the country's burgeoning tourism sector, targeting New Zealand's stunning landscapes, was overwhelming small communities, such as those at Waiheke Island, Akaroa and Wanaka. Exotic plantation forestry, which was eligible for carbon credits in recognition of its role in sequestering carbon and buoyed up by a now rising carbon price, was

Dunstan Downs Station, which runs through the Lindis Pass shown here, also went into tenure review. The initial proposal was that the land fronting the state highway would be freeholded and therefore subject to pastoral intensification, forestry or other development. After submissions by EDS and others almost all the pastoral lease at Dunstan Downs Station was brought into Crown ownership. RAEWYN PEART

starting to expand in sensitive landscapes such as Te Pataka-o-Rākaihatū Banks Peninsula. Weeds and pests were also continuing their march over the landscape countrywide.

In 2019, EDS decided to initiate a new landscape research project which was to include five case studies. Two of these, Banks Peninsula and the Waitākere Ranges, revisited places that were the subject of EDS's 2004 study. New areas investigated were the Tīkapa Moana Hauraki Gulf islands, the Mackenzie Country, and tourism and landscape protection. The findings of the case studies were integrated into a report titled *Caring for the Landscapes of Aotearoa New Zealand* released in 2021. The report also looked at the evolution of case law and landscape assessment practice, as well as revisiting a range of international protected landscape models. Lara Taylor from Landcare Research specifically documented Māori perspectives on landscape which were integrated throughout the report.

The project was led by EDS policy director Raewyn Peart, assisted by in-house lawyer

EDS's 2019 landscape study revisited some areas that were the subject of the initial 2004 landscape report, including Banks Peninsula, where Akaroa Harbour, shown here, is located.
RAEWYN PEART

RAEWYN PEART

Diane (Di) Lucas ONZM grew up on Bendigo Station in Tarras, Central Otago. She attended the University of Otago, obtaining a Bachelor of Science majoring in botany, and then Lincoln University, where she completed a Masters of Landscape Architecture in planning. Lucas began her career as a landscape architect in 1974, at the Ministry of Works and Development. In 1979 she left the ministry because 'I couldn't work under a regime that was destroying rural New Zealand' and established her own landscape architect practice, Lucas Associates Limited. Over a career spanning close to 50 years, Lucas has undertaken landscape and environmental planning assessments and related work for a wide variety of projects. Her work on the 1993 Canterbury Regional Landscape Study set the best practice standard in landscape assessment which is still used today.

Lucas sat on the Environment Council, which advised the government before the Ministry for the Environment was established, the Land Settlement Committee which administered South Canterbury pastoral leases, and the New Zealand Conservation Authority which oversees the conservation system. She has also been a regional councillor. Lucas served as chair of the Nature Heritage Fund for 27 years and for a decade was a member of Ngā Whenua Rāhui, which supports protection of indigenous ecosystems on Māori-owned land. She has served as President of the New Zealand Institute of Landscape Architects and is a life member. In 2019 Lucas was made an officer of the New Zealand Order of Merit for her services to conservation.

Lucas has long supported EDS's work, particularly on high-country landscape issues. She assisted EDS's case before the board of inquiry on the New Zealand Coastal Policy Statement and with assessing the landscape and ecological impacts of the cow cubicle proposal on Ōhau Downs Station. She also helped with EDS's tenure review proposals for Ferrintosh and Dunstan Down stations. EDS worked closely with Lucas on the proceedings associated with Plan Change 13 to the Mackenzie District Plan, where she was acting for the Mackenzie Guardians. Lucas also gave evidence in the Ruataniwha dam case and provided advice to EDS on the landscape impacts of mussel farms in Port Gore and the Blue Endeavour proposal to establish salmon farms in Cook Strait.

EDS has always sought ideas from the wider international context. Shown here are EDS policy director Raewyn Peart (left) and EDS in-house lawyer Cordelia Woodhouse in Scotland investigating national parks. GARY TAYLOR

While in the Mackenzie, Raewyn Peart and Cordelia Woodhouse met with many locals to discuss the challenges facing the area. Shown here is Woodhouse with salmon farmer Rick Ramsey looking over Lake Benmore. RAEWYN PEART

Cordelia Woodhouse. It involved considerable fieldwork; in particular, interviewing land owners and managers on the ground to understand their perspectives and the drivers that led them to adopt particular land-management practices.

It also involved an international research trip to Europe and the UK to see first-hand the landscape models operating in various countries. Woodhouse and Peart were guided around Germany, the Netherlands and Brussels by former New Zealander and Brussels resident Anton Gazenbeek. With Gary Taylor, they visited national parks and other protected areas in England and Scotland.

EDS has always sought ideas from the wider international context, has brought the good ones back to New Zealand, and has looked to build local solutions on such foundations.

The prospect of massive dairying operations in the Mackenzie Country had shocked not only EDS. The statutory agencies under whose watch it had happened were also concerned. They wanted to understand what had happened and what they could do better in the future. As a result, the Department of Conservation and LINZ agreed to support EDS in undertaking a case study of the Mackenzie Basin as part of a broader landscape protection project.

During 2019, Raewyn Peart and Cordelia Woodhouse undertook three field trips to the Mackenzie and interviewed over 40 people, including runholders, agency staff, experts and other stakeholders. In particular, the EDS researchers were keen to understand the perspectives of those currently farming the land. Much of the research was undertaken over the cold winter months but Peart and

Farming intensification was not the only pressure on the Mackenzie Basin, tourism was also burgeoning prior to Covid-19 putting local infrastructure, such as public toilets and car parks under strain. Shown here are tourists at the Church of the Good Shepherd on Lake Tekapō.
RAEWYN PEART

Woodhouse were warmly welcomed into many farm kitchens for a chat.

The case study reviewed the geological and biological history of the area, historical, current and future landscape pressures and how well the management framework was addressing them. The report concluded with a series of recommendations on how the management approach could be strengthened in the future. These included the creation of a 'Mackenzie Basin Heritage Landscape', with protections under the RMA or Conservation Act, and the establishment of a 'Mackenzie Drylands Protected Area'.

One of the biggest gaps in the statutory framework identified by EDS was the lax provisions in the Waitaki District Plan, which applied to the southern half of the basin. In this area, farming intensification was still a permitted activity. A review of the district plan had been underway since 2014 but has yet to result in tighter provisions being notified.

Although the EDS report on the Mackenzie Basin was critical of agency performance, it was surprisingly well received. Then Environment Canterbury Acting Chief

Cordelia Woodhouse graduated from Victoria University with a Bachelor of Laws and Bachelor of Science. She joined EDS as a solicitor in November 2018. Woodhouse managed EDS's litigation programme and she was involved in the Simons Pass Station case and tenure review proposals in the Mackenzie Basin, the review of the Marlborough Sounds Environment Plan and the Mataura ouvea premix case. She also managed appeals arising from the Otago regional policy statement, including those related to Port Otago and OceanaGold. In addition, Woodhouse was a key researcher for the EDS landscape project.

As Woodhouse reflected, 'policy work is getting down there on the land and talking to people, as well as the nitty-gritty stuff of going through the plans, and finding out the legal side of the issue. The landscape project was a pretty big highlight, in particular the Mackenzie Basin and Banks Peninsula work. In those places there was more of a story to tell. The responses to both case studies were quite incredible. Even with the Mackenzie Basin case study, where we were quite critical of councils and DOC, we had a good response. The Banks Peninsula report was also very highly regarded.'

Woodhouse moved on to a new role as a solicitor at Ellis Gould in late 2022.

Executive Stefanie Rixecker (later Chief Executive) said on behalf of the joint agencies, 'we acknowledge there is much more to do and the case study provides innovative ideas, some of which we're already pursuing and others we need to consider further'.

On 5 September 2020, Conservation Minister Eugenie Sage announced the creation of the Tū Te Rakiwhānoa Drylands in partnership with mana whenua. As part of the announcement, five pieces of Crown land totalling around 11,800 hectares were added to existing conservation land and private covenants to make a total area of 31,300 hectares of legally protected land on the Mackenzie Basin floor. It was a good start but somewhat smaller than the 100,000 hectares of protected area envisaged in the Mackenzie Agreement.

Over recent times, the impetus for more protection seems to have diminished and the agencies have lost focus. More work is clearly needed to protect the Mackenzie Country.

In both the 2004 and 2021 landscape reports, EDS promoted the idea of a third way, where there would be an extra layer of protection over private land for areas of outstanding landscape values. Although commonplace overseas, this idea has still to get traction in New Zealand. While some places have been protected, important landscape values continue to be lost.

13. Cleaning up rivers

BY THE TIME EDS was re-established in 1999, concerns around freshwater quality had moved on from piped sources of pollution (such as the sewage in EDS's early Huntly case), which had largely been cleaned up under the RMA, to diffuse sources of pollution — particularly from intensive dairying.

The proliferation of dairy cows on the Canterbury Plains created nitrogen pollution problems for nearby rivers, including the Waimakariri shown here. RAEWYN PEART

FROM THE 1990s dairy cows were proliferating around the country, with herds in Canterbury increasing from just 113,000 in 1990 to 1.2 million in 2019, and in Southland from just 38,000 to 636,000 over the same period. And it was not just the total numbers of cows that were increasing but the numbers of cows per hectare.

This significant intensification of land use required enormous amounts of synthetic nitrogen fertiliser, along with irrigation water on the dry plains, to grow the grass required to feed the burgeoning number of cows. Excess nitrates from fertiliser and cow urine started draining into lakes and rivers, turning their waters green from excess weed and algal growth. Nitrates also drained into groundwater, contaminating rural water supplies.

The issue of water pollution from intensive dairying first hit the public consciousness in the early 2000s after Fish and Game launched a 'Dirty Dairying' campaign. The organisation coined this term to clearly link dairy farming with environmental harm. It also sought to highlight the abject failure of regional councils to deal with the issue and of central government to hold them to account.

The term 'dirty dairying' resonated with the public and went viral. It was a stark contrast to the 'clean and green' image New Zealand had prided itself on and the 'Pure New Zealand'

Previous pages: Waikato River near Huntly. RAEWYN PEART

Fish and Game ran a very effective 'dirty dairying' campaign, which highlighted the water pollution caused by intensive farming. Shown here are cattle in a nutrient-enriched waterway adjacent to Lake Ellesmere.
ROB SUISTED/NATURESPIC

tagline the government had launched in 1999 to attract tourists. Many people had seen their favourite rivers deteriorate to a point where they were no longer swimmable or fishable. Dirty dairying seemed to clearly pinpoint the nub of the problem.

The public upwelling of concern meant that the government could no longer ignore the issue. But it was dealing with a sector that was of huge economic importance to the country. Dairy exports were booming, particularly milk powder, providing badly needed foreign exchange. Those in power didn't want to kill the economic 'golden goose'.

It was in this climate that, in 2003, the Labour Government initiated the 'Sustainable Water Programme of Action' under then Environment Minister and former teacher Marian Hobbs. The initiative quickly became bogged down and was soon known as the 'Programme of Inaction', due to its slow progress and lack of any tangible outcomes.

In 2006, a draft national policy statement on freshwater was finally produced and a board of inquiry established to scrutinise it. But then the process appeared to stall.

In October 2005, after the general election which retained the Labour-led Government, former teacher David Benson-Pope replaced Hobbs as Minister for the Environment. This was after losing his ministerial positions in the lead-up to the election due to allegations of misconduct during his teaching career.

Benson-Pope was embroiled in further

Lake Taupō was subject to a novel approach to reducing nitrogen discharges from the surrounding catchments when a cap-and-trade system was introduced by the Waikato Regional Council. EDS strongly supported the approach but sought a more robust design. RAEWYN PEART

political controversy leading to him resigning his ministerial posts in July 2007. Former lawyer David Parker then became acting Environment Minister for a few months before former teacher Trevor Mallard was appointed minister. Mallard held the portfolio for just over a year, until the general election, which Labour lost.

This churn in environment ministers, with five different incumbents during the five years since the Programme of Action had been launched, no doubt contributed to the lack of traction on freshwater and other environmental issues during this period.

One of the few positive developments in the freshwater space during this time was the proposal by the Waikato Regional Council to establish a cap-and-trade system to manage nitrogen entering Lake Taupō. There were concerns about the decline in the lake's water quality due to excess nitrogen, with 93 per cent of human-generated nitrogen entering the lake sourced from pastoral farmland.

The 2005 proposal was ground-breaking. A tradeable permit system for diffuse nutrient discharges had not been previously applied anywhere in the world. EDS became involved, with law professor Barry Barton taking up the cause on behalf of the society. EDS made submissions strongly supporting the initiative but seeking more robust design, including flexibility to adjust the cap. The scheme became fully operative in 2011 and has helped maintain high water quality in the lake. Elsewhere, water quality was rapidly deteriorating.

WHEN FACED WITH a government that was achieving little to improve environmental protection, EDS looked for other ways to make progress. One of these was to work with other protagonists to see if those from a wide range of perspectives could agree on a way forward. If Māori, business and the environmental sector were in agreement, it seemed more likely that government would follow behind. The political risk would be taken out of the issues and, in this way, civil society would lead reform. This was to mark a significant shift in approach by EDS.

In June 2008, EDS convened its annual conference in Auckland. That year the conference had a focus on the environmental impacts of rural activities and was titled 'Conflict in Paradise: The Transformation of Rural New Zealand'.

The issue of farm pollution had been growing in public prominence, supported by Fish and Game's 'Dirty Dairying' campaign. According to the Lincoln University bi-annual survey on the public perception of the state of the environment, by 2008 farming was for the first time seen by the public as a bigger culprit in causing water pollution than sewage and stormwater runoff.

Reflecting collective frustration over the lack of government action, the conference culminated in an agreement facilitated by EDS to establish a Sustainable Land Use Forum. This would be a collaborative group that would work across sectors to find ways of making New Zealand agriculture more sustainable. Those who agreed to take part with EDS included Federated Farmers, Fonterra, Lincoln University, University of Auckland, the Ministry of Agriculture and Forestry, the Ministry for the Environment, Ecologic, Fish and Game and Forest and Bird. Others invited to join included Local Government New Zealand and iwi authorities.

The forum aimed to find common ground on what new water policies, in particular, should be developed. If government was failing to act, the stakeholders would get together to see if they could agree on a way forward.

The formation of the Sustainable Land Use Forum was inspired and guided by environmental campaigner Guy Salmon's research into the operation of collaborative processes in Scandinavia. The Maruia Society, which Salmon had headed since its establishment through the merger of the Native Forest Action Council and EDS in 1988, had morphed into the Ecologic Foundation in 1999 (co-incidentally the same year EDS was re-established as a separate entity). Salmon was associated with the Bluegreens in the National Party and he persuaded his colleagues there to support the collaborative approach, which became a plank in the National Party's 2008 election manifesto.

A year later, at the June 2009 EDS conference, 'Reform in Paradise: Threat or Opportunity', Environment Minister Nick Smith announced that the government would back the initiative. By that time, 45 organisations had been brought into the process. The government subsequently provided funding to the renamed Land and Water Forum.

Business journalist Rod Oram described Minister Smith's announcement in a *Sunday Star Times* article as a 'brave and worthy' one. He saw it as 'a radical departure from the dysfunctional adversarial processes we've

Prior to becoming a director of EDS, Alastair Bisley chaired the Land and Water Forum which against all odds managed to reach agreement on a set of 53 recommendations. Bisley is shown here presenting on the Land and Water Forum at the 2013 EDS conference. RAEWYN PEART

long applied to complex problems'. However, 'making the Land and Water Forum work will be extremely challenging,' said Oram. This was because 'the 45 or so stakeholders around the table have very divergent, strong views on ownership, management and use of water'.

Alastair Bisley, who chaired the Land and Water Forum (and later became an EDS director), recalls that when the idea of a Scandinavian-style collaborative process came up, people thought it had to be better than spending their time in court.

> The more they have done it [collaboration], the more they have felt comfortable doing it, and the more they felt they could trust one another and actually make progress. What we started doing was to talk about experiences. We asked everyone to talk about what their overall strategy was for their enterprises, how that impinged on water and what they were doing in relation to water.

After more than a year's deliberations, and against all odds, the forum nailed a set of 53 agreed recommendations in its first report, which it presented to Minister Smith in September 2010. The recommendations included a proposed Land and Water Commission to provide oversight of freshwater management, beefing up the Parliamentary Commissioner for the Environment to provide independent monitoring, using audited self-management at the farm level, providing clear national policies, objectives and standards, and giving more support to regional councils.

The forum also agreed that regional councils should prepare water plans using an inclusive and collaborative process and that there should be a nationally funded effort to clean up degraded water bodies. In addition, it urgently wanted to see a national policy statement on freshwater.

The forum's recommendations paved the way for New Zealand's first national policy statement on freshwater management in 2011. The forum had been pivotal in breaking the 'policy deadlock' that had been in place since the mid-1990s. But its work was not done, and the forum would go on to produce three more reports as well as additional advice to ministers.

Many land uses have an impact on water quality, including forestry harvesting, agriculture and horticulture, shown here in Marlborough District. The Land and Water Forum sought to come up with a framework that would help clean up water pollution. RAEWYN PEART

Its subsequent recommendations included setting environmental limits for freshwater, specifying the minimum acceptable state, and managing within them. The recommendations influenced a further iteration of the freshwater national policy in 2014, which introduced a national objectives framework. And, in 2017, the framework introduced the Māori concept of te mana o te wai (the integrated and holistic wellbeing of a freshwater body).

Natasha Garvan was involved in the Land and Water Forum on behalf of EDS during its early stages.

> I remember vividly one of the very first meetings that Gary [Taylor] took me to. When we broke for morning tea the environmentalists went to one corner and the agricultural players went into another corner and talked amongst themselves. Fast forward three years and I was called to another small group meeting in Wellington that Gary couldn't attend. When I arrived, I saw the guy from Dairy NZ give Kevin Hackwell from Forest and Bird a big hug. Then they walked in greeting one another. That epitomised how far it had come.

But the work of the forum was not uncontroversial. Fish and Game, Forest and Bird and Federated Mountain Clubs all withdrew at various stages due to concerns about the government's slow progress in developing robust freshwater policy. Freshwater ecologist Dr Mike Joy repeatedly raised concerns about the ongoing deterioration of freshwater quality, questioning the country's clean and green credentials, and highlighting the role of powerful agricultural industry lobby groups in obstructing change.

By 2018 the limits of a collaborative approach were being reached. The Land and Water Forum had now been operating for the best part of 10 years and was running up against controversial issues on which reaching consensus was increasingly difficult. These issues included water allocation and the management of nitrogen, a matter which went to the very heart of the dairy production system. Also, although the forum had positively influenced the development of freshwater policy, many of its recommendations had not been taken up by government. In July 2018, the forum put itself into abeyance.

A Labour-led Government had been elected to office in October 2017 on the back of promising to clean up the country's rivers. Freshwater management had for the first time become an election issue, with reports showing continued degradation of the country's lakes and rivers. Environment Minister David Parker took a more directive role in addressing the issue. It was time for the government to face up to some hard decisions.

As Taylor recalled,

> the Land and Water Forum was a long, challenging and ultimately rewarding process. The forum members collectively held out for progress across all aspects of freshwater management. Initial enemies became almost friends — certainly mutual respect across sectors improved. Minister Nick Smith, often unfairly criticised for a lack of response to our recommendations, had a hard job getting some of his cabinet colleagues to embrace reform. With a more progressive cabinet we might have made more progress more quickly.

Nick Smith, the Environment Minister who supported the initial establishment of the Land and Water Forum, observed,

> EDS's biggest contribution to the freshwater issue was in the establishment of the LAWF [Land and Water Forum]. Relationships under the Clark Government's water programme of action had become so acrimonious that there had been very little progress. EDS took the view that New Zealand could not make progress on freshwater unless there was active dialogue with both the farming and hydro electricity sector.
>
> The LAWF proved pivotal in delivering New Zealand's first national policy statement on freshwater in 2011. It was not perfect, but it was a huge step forward to have a national framework. Not only did it shift the legal obligations on to regional councils, but it also hugely shifted the mood of the rural community, from one of denying that there was a problem to one of accepting there was a problem and engaging on what sort of solutions would work.
>
> It's unfinished work. Improvement of New Zealand's freshwater is going to be a generational journey. But I think people will look back on the establishment of the LAWF and the first NPS [national policy statement] as a pivotal turning point in New Zealand's management of freshwater.

Natasha Garvan graduated with a Bachelor of Laws (Hons) and a Bachelor of Arts in political studies from the University of Otago in 2008, after which she joined Bell Gully as a lawyer. She subsequently completed a Bachelor of Arts (Hons) in political studies from the University of Auckland.

Garvan was seconded from Bell Gully in late 2009 to work for EDS as an in-house lawyer. She worked part-time for EDS over the next four years. Garvan ran the community advice service and the EDS litigation programme. She was involved in the Kaipara 'blank pages' case, the cubicle cow case in the Mackenzie Basin and the King Salmon case at board of inquiry level. She was also co-author of the EDS freshwater community guide.

Garvan observed, 'I really liked how inclusive the EDS team always was. When they had strategic retreats they would still include me as part of the team, across at Rotoroa Island or wherever it was. Gary also flew me over to Sydney for a climate change and business conference meeting. He always fully involved me in the opportunities that were available. That gave me a sense of ownership of the work we were doing and a feeling we were contributing to something bigger than ourselves.

'The absolute highlight for me was the Land and Water Forum governance working group which I was on. We came up with the idea of an appeal structure, where there wouldn't be appeals to the Environment Court if the council agreed with the collaborative group. That was ultimately adopted for the Auckland Unitary Plan and we came up with the idea in the Land and Water Forum. It was pretty exciting as a junior lawyer to be involved and to meet all of those important people.

'I can honestly say that I definitely would not be the lawyer I am today, or where I am at, if not for my experience at EDS. I wouldn't have had anywhere near the same exposure or learning opportunities.'

In 2015, Garvan started Food, Farms and Freshwater (3F) alongside environmental consultant Rhys Millar and farmer Mike Barton, designed to create a pathway to enhancing the environment whilst adding value to the New Zealand economy. In 2019 Garvan became a partner at Bell Gully, specialising in environmental and resource management law.

Although the Land and Water Forum has been disbanded, EDS has continued to influence the ongoing development of freshwater policy. Gary Taylor was a member of the Freshwater Leaders Group that helped develop the 2020 package of freshwater reforms. These considerably strengthened freshwater policy, establishing a hierarchy of obligations which put the well-being of waterbodies first, and expanding and strengthening the national objectives framework amongst other things. Taylor now sits on the Freshwater Implementation Group, which is overseeing the roll out of the policy.

As well as seeking to strengthen freshwater policy at a national level, EDS was also keeping an eye on what was happening on the ground. Successive governments had adopted a policy of supporting irrigation to increase productivity in the primary sector, including through providing public funds, and this was prompting renewed interest in dam building.

CENTRAL PLAINS WATER was an ambitious project that sought to irrigate much of the central Canterbury Plains by taking water from the Waimakariri and Rakaia rivers. Promoted by the farmer-owned Central Plains Water Limited, the proposal included a 55-metre-high dam and a 280-million-cubic metre storage reservoir in the Waianiwaniwa Valley, located in an area called Malvern Hills. There was to be a headrace canal some 53 kilometres long and 460 kilometres of water races connecting the reservoir to farms out on the plains.

The dam face was to be just 1.5 kilometres away from the small town of Coalgate. In total, the project was to irrigate some 60,000 hectares of farmland. It was strongly supported by the Selwyn District and Christchurch City councils.

What was somewhat audacious about the plan was that a group of Canterbury Plains farming interests had formed a private company which then obtained 'requiring authority' status from Environment Minister David Benson-Pope in 2005. A large stakeholder in the farmers' venture was Dairy Holdings Limited, which operated 57 dairy farms. Once the requiring status was approved, the privately owned company was then able to place a notice of requirement over other farmers' properties in Malvern Hills in order to flood and build a dam and canal infrastructure over the land.

The notice of requirement was served in 2006 and meant that the Malvern Hills landowners were unable to use their land for anything that might 'prevent or hinder' the project. In addition, the requiring authority status came with the power to forcibly acquire land under the Public Works Act if the project went ahead.

The ability of this group of private interests to obtain such draconian powers was the result of very broad provisions in the RMA which, reflecting the privatisation of some public services such as electricity generation and transmission during the 1980s, enabled private providers to gain the same powers as were ordinarily only exercised by public bodies to protect land for public infrastructure, such as motorways and water pipelines.

Malvern Hills is an old coal-mining and farming area on the edge of the Canterbury Plains to the west of Ōtautahi Christchurch. The Malvern Hills Protection Society, comprising farmers and local residents concerned about the Central Plains Water proposal, was formed in 2001. By the time the group approached EDS for assistance in 2007 a lot of water had gone under the bridge, so to speak.

The protection society had unsuccessfully gone to the Environment Court to challenge the decision of the Environment Minister granting requiring authority status to the irrigation company. It had not gone well; when reaching its decision in August 2007 the court had found the group's case to be 'frivolous, vexatious' and 'otherwise an abuse of process'. Costs were awarded against it.

A large suite of applications had been lodged with the Canterbury Regional Council and Selwyn District Council between June 2005 and March 2007 in order to consent the proposal for the Central Plains Water scheme. These applications generated well over 3000 submissions, including those by the Malvern Hills Protection Society and numerous individual members of the society affected by the scheme.

The Malvern Hills Protection Society approached EDS for legal assistance after unsuccessfully going to court in 2007. EDS helped the society to prepare their case against the Colgate dam, which would have flooded land in the Malvern Hills. Society members prepare for a festival to raise awareness of the Central Plains Water proposal. COURTESY OF ROSALIE SNOYINK

Rosalie Snoyink, a driving force behind the Malvern Hills Protection Society, recalls first hearing about EDS in late 2007 when an Auckland lawyer who was unable to act for the group due to a conflict of interest suggested she contact Gary Taylor. As she recalls, 'I remember Gary flying down and I took him on a site visit. He gave it some thought and agreed to help us. That was the beginning of a very good relationship for both groups.'

The hearing of the applications by a group of four independent commissioners chaired by environmental lawyer Philip Milne commenced on 25 February 2008. It was an extremely long hearing, taking 68 days over a period of two and a half years and including five site visits. One of the visits was to the Malvern Hills.

The evidence presented was voluminous, including that from 37 witnesses presented by the applicant Central Plains Water alone. It was more than enough to completely swamp any small community group. Three members of the Malvern Hills Protection Society committee — Maureen Robertson, Liz Weir and Rosalie Snoyink — attended every day of the hearing, their presence often acknowledged by the hearing commissioners.

EDS arranged for lawyer John Burns to take on the case with assistance from EDS in-house lawyer Shay Schlaepfer. EDS director and former water engineer Garry Law helped out with reviewing material. In addition to securing expert evidence to support the case, the EDS team helped prepare numerous briefs of evidence from landowners and residents in the Malvern Hills area.

As Rosalie Snoyink recalls,

> when we went to the hearing and gave evidence it was an emotional day. There had been a huge amount of work to get all the lay submissions completed and in a presentable form. EDS had sent us some key dates, had given us some guidelines and told us to get lots of photos into the submissions.

Some key people were very reluctant at the beginning. They were so emotionally involved in the future of their farms and their land was going under water. But we got there in the end.

When it came to the hearing it took something like two days for the Malvern Hills submitters to be heard. Up until that time, I thought it was important to keep emotion out of RMA processes, to just focus on the facts. But I think it was those two days where the local community finally had their say that won the case. It was very moving and everyone was so well prepared thanks to EDS.

Shay Schlaepfer: 'it was very harrowing going down to the hearings. Around 40 locals all had individual statements which we reviewed for them. They were all having sections of their land being taken for irrigation canals and their whole livelihoods being upended. They were long-standing residents with family ties to the area. It was an emotionally charged hearing.'

John Burns remembers the case well.

The Malvern Hills Protection Society which opposed the proposal was made up of the most disparate bunch of people I have struck. They ranged from high-country farmers who hadn't been out of the Malvern Hills for years, through to lowland Canterbury Plains farmers and residents of the village of Coalgate, which was going to be right beside the dam. The people in Malvern Hills were concerned that their farms were going to be compulsorily taken and flooded for the dam. The people in Coalgate were concerned about having a large dam so close to their town and that if the dam burst they would be flooded.

There were some business owners on the plains who were affected and some farmers on the plains who didn't want to enter the scheme. Then there were the green environmentalists who were opposed to the effects on the two rivers of taking water, and a jet boat operator on one of the rivers worried about losing his business.

It was a group of people who would not normally be aligned. Rosalie did a wonderful job of bringing them all together and co-ordinating the case. In the end, we decided not to get involved in the water issues and ran the case pretty much on the social and economic effects.

I remember a wonderful brief of evidence from a young farmer, who farmed in the foothills of the Malvern Hills. His farm was going to be compulsorily taken and flooded for the dam. He explained how his children often woke at night with nightmares about their farm being covered in floodwater. They had heard the adults talking about the scheme and thought one day their farm would flood.

A second thing he said that remains vivid in my memory is that he had bought the farm deliberately in the foothills, and not the plains, because he knew there was plenty of natural rainwater there. It was naturally irrigated. So he was incensed that there were proposals to take his perfectly viable naturally irrigated farm to flood it and use it to irrigate the plains which were inherently unsuitable for dairying as they had no water. His evidence was very telling.

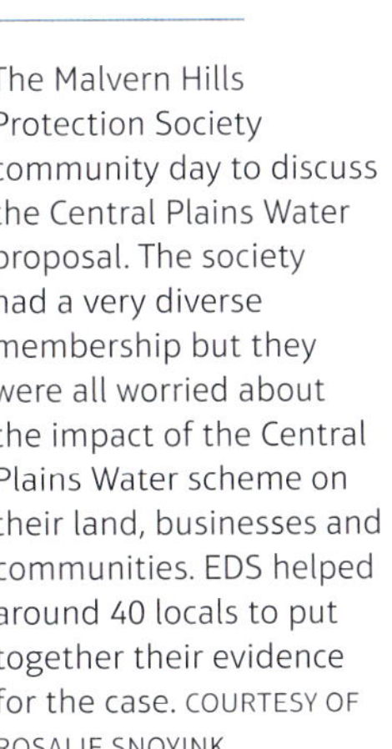

The Malvern Hills Protection Society community day to discuss the Central Plains Water proposal. The society had a very diverse membership but they were all worried about the impact of the Central Plains Water scheme on their land, businesses and communities. EDS helped around 40 locals to put together their evidence for the case. COURTESY OF ROSALIE SNOYINK

In addition to the briefs from local landowners and residents, EDS called several expert witnesses. They included landscape architect Di Lucas, who gave evidence on the landscape impacts of the dam and reservoir, economist Tim Hazledine and University of Canterbury natural hazard expert Jocelyn Campbell.

As John Burns recalled,

> one of the components of the case was the dam to be built near the village of Coalgate and what would happen to the residents if it burst. We found a senior scientist at Canterbury University who was retiring, Jocelyn Campbell. She didn't want to give evidence, as she had to get her life's work out of her office over the following few weeks, so was short of time. But we persuaded her to do so.
>
> She gave evidence that there was a fault line in the area which either ran through or centred on Darfield. If there was an earthquake she thought there was a good chance that the dam would break and the town would be flooded.
>
> The other side had produced all manner of experts to say that wasn't the case and the issue got lost in a morass of counter evidence. But a few years later the Christchurch earthquake happened [in 2010 the first Christchurch earthquake occurred, sometimes referred to as the Darfield earthquake]. Everyone seemed surprised that the earthquake had hit Darfield but she knew the fault was there and that it could go then or in a hundred years' time.

It was extremely fortunate that the dam had not been built.

In the end, the commissioners decided to decline the dam and reservoir, and the irrigation project was re-designed to operate as a run of river scheme where water is taken from the flowing rivers without any storage capacity.

RAEWYN PEART

John Burns graduated with a Bachelor of Laws from Victoria University of Wellington in 1969. He moved to Auckland in 1985 to specialise in planning and water law. Burns worked for 25 years as a contracted in-house solicitor for the Auckland Regional Council. He also maintained his own legal practice, which from 2010 became Burns Fraser. In 2013 Burns became a consultant at K3 Legal, specialising in resource management and local government law. In 2018 he retired from legal practice.

Burns has taken on many cases for EDS, including the Central Plains Water case, the Horizons One Plan litigation and the Poronui Trust and Sade Development cases in Taupō. He noted, 'EDS has achieved a lot through its litigation programme. There is no doubt that EDS's involvement in litigation got the Taupō District Council to put its plan in shape. One thing the individual cases do is highlight to the council and the court the deficiencies in the plan which they, with luck, remedy. Litigation is also of incalculable value to community groups. None of those groups could have run cases without EDS support.

'I have found that EDS's reputation is very good. It has a reputation for being quite balanced and sensible, not outrageously controversial. It's clearly respected in the courts. I suppose you could say EDS is well regarded by the establishment without being part of the establishment.'

As Snoyink reflected,

> we won the case because we were represented by John Burns. He taught us how to write and present submissions. From this case we learned a lot. We learned about the effectiveness of community participation in the decision-making process. A compromise was reached, and the decision was not appealed by either side. It showed that power does not necessarily lie with corporates or politicians, it is just as strong within citizenry.

—000—

A SECOND LARGE and ambitious irrigation scheme that caught EDS's attention was the one proposed for the Tukituki River in Hawke's Bay. It included a large irrigation dam to be built in the foothills of the Ruahine Ranges, which would impound 90 million cubic metres of water on the Makaroro River, a tributary of the Tukituki River. If built, the 83-metre-high dam would be the largest irrigation dam in the country and it would create a 7-kilometre-long storage lake flooding conservation land.

The dam was to be accompanied by an irrigation network to water 25,000 hectares of land in order to facilitate the growth of dairying and horticulture within the catchment. The impacts of the project would be big and at a landscape scale.

The scheme required a change to the

Hawke's Bay Regional Plan (Plan Change 6), 17 resource consents for the water-storage scheme, and a notice of requirement to designate land. The bundle of applications lodged by the Hawke's Bay Regional Council (for the plan change) and Hawke's Bay Regional Investment Company (for the resource consents and notice of requirement for a designation) was 'called in' by Environment Minister Amy Adam and Conservation Minister Nick Smith in June 2013. This meant they would not be considered by councils, which was the normal procedure, but by a ministerially appointed board of inquiry.

The scheme was costed at $275 million, with on-farm costs covered by the farmers themselves bringing the total to $650 million. The regional council had committed $80 million.

The Tukituki River, shown here running through Heretaunga Hastings District, was the subject of a proposal to construct an enormous irrigation dam on one of its tributaries which would have created a 7-kilometre-long storage lake. DOUGAL TOWNSEND, GNS SCIENCE

The scheme was highly contentious and 384 submissions were lodged. Concerns focused not only on the environmental impacts of the dam itself, but also on the flow-on effects of the intensified dairying and horticulture it would enable, including impacts on river health.

Freshwater in the middle and lower parts of the catchment was already in a degraded condition with extensive mats of algae and slime evident on the bed of the Tukituki River. But that did not deter farmers, with departing Federated Farmers chief executive Conor English being reported as stating in July 2014 that Hawke's Bay would be doomed to Third World country status if the Ruataniwha dam was not built.

The Department of Conservation had initially drafted a detailed 50-page submission opposing the project, but this was pulled at the last moment and a neutral submission of only a few lines was lodged. In fact, the department was entirely absent from the hearing. There were rumours that it had received a political directive not to voice its opposition to the plan.

By this time, the National-led Government was into its second term and was strongly pursuing an economic development agenda. The absence of the Department of Conservation meant that the burden of testing the applicant's arguments fell on the voluntary sector. Doing this effectively was exceedingly challenging.

EDS was represented at the hearing by lawyer Rob Enright, now an independent barrister, and in-house lawyer Nicola de Wit. They worked closely with lawyers acting for Forest and Bird and Fish and Game, with both those organisations also submitting against the proposal.

The applicants had clearly spent a long time preparing their case. They produced an enormous quantity of highly technical reports and evidence to which submitters, including EDS, had only a few weeks to respond. The thousands of pages of material were overwhelming. At the same time, EDS's experts were not given access to the actual dam site. Gary Taylor described this as 'deeply prejudicial and the process fundamentally flawed'.

At the beginning of the hearing the board of inquiry had 5329 pages of reports and other documentation before it; submissions (including the summary of submissions) running into 6053 pages; and 181 briefs of evidence from 131 witnesses totalling 6853 pages. And this evidently didn't cover the ground adequately, because during the hearing a further 6326 pages of documentation were produced (including 90 exhibits). By the time the hearing concluded, the board had before it more than 28,000 pages of material. By any measure this was technical overkill.

The hearing began on 18 November 2013 and concluded the following year on 21 January. One of the key issues in contention, and one that EDS focused on, was how Plan Change 6 proposed to manage phosphorus and nitrogen discharges into the river. The proposed plan change focused on managing nuisance growth of algae and other material on the riverbed (called periphyton), rather than the broader health of the river ecosystem, and sought to do this through adopting a 'single nutrient' approach. This would control phosphorus discharges into the river water but adopt a 'hands off' approach when it came to nitrogen.

The single-nutrient approach was based on the scientific understanding that periphyton growth requires both phosphorus and nitrogen to be present in the water, so that by constraining one of the elements (in this case phosphorus), increasing levels of the other (nitrogen) would not necessarily result in increased periphyton.

This implied that so long as phosphorus was controlled, nitrogen levels could be increased in an unconstrained manner, so long as they did not reach a level that was 'toxic' to river life and killed fish and other creatures outright. But as Dr Russell Death, a freshwater ecologist who gave evidence on behalf of the Hawke's Bay and Eastern Fish and Game Council, explained to the board, it is similar to the effect of alcohol. 'Long before a person consumes so much alcohol that he or she is poisoned, that person will usually have become very unhealthy.'

This single-nutrient approach was strongly

The Tukituki River water-storage scheme was highly controversial and when the Department of Conservation failed to make an appearance at the hearing, the heavy lifting was left to environmental groups such as EDS, Forest and Bird and Fish and Game. The mouth of the river is shown here in flood. ROB SUISTED/NATURESPIC

supported by those in the nitrogen-producing primary sector, including Federated Farmers, Dairy NZ and the Fertiliser Association of New Zealand. There was good reason for this. Phosphorus is much easier to control in dairy farming than nitrogen. Both nitrogen and phosphorus are sourced from fertiliser application, external feed and cow urine and dung. But nitrates (a very water-soluble form of nitrogen) readily leach through the soil into groundwater, which then drains into rivers.

By far the largest source of nitrogen in the soil comes from cow urine patches. It is therefore difficult to control nitrogen leaching from a farm without reducing stock numbers or housing all the animals indoors. This means the intensification of dairying facilitated by the proposed irrigation scheme would inevitably result in increased nitrogen leaching into the Tukituki River. Phosphorus, on the other hand, binds to the soil, so measures such as riparian fencing and planting which reduce soil erosion into rivers can help manage its loss while maintaining intensified farming.

EDS, alongside Fish and Game and Forest and Bird, supported a 'dual-nutrient' approach. This was based on the argument that managing only phosphorus was exceedingly risky. As Nicola de Wit recalls,

> there was some really interesting science. I remember we had freshwater expert Kate McArthur giving evidence for EDS and there was a really interesting relationship between nitrogen and phosphorus and how you could control that ratio to influence outcomes in the system.
>
> The applicants were relying on a tricky approach where if you kept one low it didn't matter what the other one did. But Kate and others were of the opinion that this was very risky. It was a very strict engineering approach, but if you get it wrong you are in big trouble. And it would be 100 years before you would start to see if you had got it wrong. So if you backed the wrong horse, you wouldn't see it for a long time and the damage would be entrenched in the system.

It came down to risk and precaution.

In the end, EDS and the other environmental groups won on the issue, with the board of inquiry favouring a dual-nutrient control approach which managed both phosphorus and nitrogen. Although the board approved the plan change and resource consents, it imposed an inorganic nitrogen limit of 0.8 mg/l across most of the catchment. This was set at a level designed to protect ecological health. In the board's view, it was also more likely to

Environmental groups successfully argued for a dual-nutrient approach to managing water quality in the Tukituki catchment, but this was undermined by a last-minute change to the board of inquiry's decision after an additional submission by the Hawke's Bay Regional Council, one of the proponents of the water storage scheme. RAEWYN PEART

give effect to the National Policy Statement on Freshwater Management which had come into force in 2011 after the efforts of the Land and Water Forum. The level was a significant reduction on the 2.4 to 3.8 mg/l limits proposed by Hawke's Bay Regional Council, which were three to four times larger, and based on a toxicity approach.

Setting the 0.8 mg/l nitrogen limit was very significant, as it was already exceeded in large parts of the Tukituki catchment, with some places showing levels double the limit. This meant that rather than intensifying land use through irrigation, which was what the dam was intended to achieve, the regional council would need to reduce nitrogen leaching into the waterways. And this was likely to necessitate less application of fertiliser and fewer cows. Overall, the decision meant that the commercial viability of the dam and irrigation project was in question, and it seemed unlikely it would go ahead.

However, there were some concerning changes to the board's decision between its draft and final reports when it came to how the nitrogen limit would apply. The wording of the board's draft report meant that individual farmers would require resource consents if they caused or contributed to nitrogen entering the catchment in excess of the 0.8 mg/l limit. However, when the final decision was issued, this had been changed. Individual farmers were to be allocated a maximum allowable nutrient loss based on the amount of nitrogen that would be expected to leach from an optimally managed fully-stocked farm on similar soil types.

This allocation had no linkage to the 0.8 mg/l limit, but overall that limit was deemed to have been complied with. This effectively created a 'factual fiction' where the limit, in fact, could be breached. The impact of the change was not minor. It would allow around 615 farms to discharge nitrogen into the waterways in excess of the limit.

It turned out that the change had been made in response to an additional submission made by the Hawke's Bay Regional Council which, wishing to support agricultural production in the region, was seeking a more lenient on-farm nutrient regime. Other parties had not been given the opportunity to comment on the submission before the final decision was issued. EDS thought this was patently unfair and so, alongside Forest and Bird and Fish and Game, the society appealed to the High Court both on the fairness issue and also on the legality of the change itself. It alleged that the change did not comply with the requirements under the relevant legislation, that is the RMA.

In its decision issued in December 2014, the High Court found that the principles of natural justice required the board to provide the affected parties with an opportunity to comment on material changes to its decision, which it had not done. It also found that the actual provisions, which deemed the nitrogen limit to be met when in fact it could be breached, were contrary to the law. The matter was remitted back to the board for reconsideration.

By the time the board issued its final decision in June 2015 investors had started to go cold on the irrigation scheme. In March 2014 Trustpower, which had been pegged to contribute up to $60 million towards the construction costs, announced that it was withdrawing as an investor on the basis that 'it would not be possible to invest in the Ruataniwha dam within its risk and return framework'. This was followed by the withdrawal of Ngāi Tahu Holdings, which had been considering investing some $50 million in the scheme, due to not having the 'right investment partner' after the Trustpower withdrawal.

The Hawke's Bay Regional Investment Company, which had been the main proponent of the scheme, subsequently indicated that it was in talks with a significant institutional investor, with rumours that it might be the Accident Compensation Corporation or Superfund, but nothing came of it in the end.

There were still some last-ditch efforts to keep the struggling project alive. The scheme promotors needed to acquire the area of land within the Ruahine Forest Park which was to be flooded. This was not a straightforward proposition as the area had very high conservation values.

The Hawke's Bay Regional Investment Company initially looked at applying for a concession to build a dam within the park, but with the high conservation values at stake, it was clear that approval would not be forthcoming. The company then looked for an alternative proposition and decided upon a land swap. It entered a conditional agreement to buy 170 hectares of nearby private land which it offered to the Department of Conservation in exchange for the 22 hectares of conservation land that would be inundated. Director-General Lou Sanson approved the swap in October 2015 on the basis that it would provide a net gain for conservation.

Forest and Bird thought this was not right and appealed Sanson's decision to the High Court. It turned on the specific provisions in the Conservation Act which prevented land designated as conservation park, which was the status of the land in question, from being directly swapped. Instead, first the conservation status had to be revoked. Forest and Bird argued that when that decision was made, it could not take into account the conservation impacts of any subsequent swap of the land. The High Court dismissed this argument but Forest and Bird won in the Court of Appeal.

The Hawke's Bay Regional Investment Company had too much at stake and decided to take it to the highest court in the land, the Supreme Court. If it couldn't get access to the conservation land, the scheme was definitely dead. But in a July 2017 decision, the Supreme Court upheld the Court of Appeal and the land swap was off the table. The Hawke's Bay Regional Council wrote off $14 million spent on the failed scheme in August 2017. But in a response redolent of those to earlier court losses by government (such as the Clyde Dam), Conservation Minister Maggie Barry indicated that government might change the law to enable such land swaps to take place.

Controversies over nutrient-leaching by farms were now erupting all over the country as the freshwater reforms were starting to bite. EDS was keen to ensure that the legal limits were being adhered to.

THE MANAWATŪ-WHANGANUI Regional Policy Statement and Regional Plan, known as the 'One Plan' become operative in December 2014. The plan applies to much of the lower North Island, extending from an area to the west of Lake Taupō, past Whanganui and down to Levin south of Palmerston North. It includes the fertile Manawatū Plains.

In June 2013, while the plan was being finalised, the Manawatū-Whanganui Regional Council (known as Horizons Regional Council) had passed a resolution stating that it would

RAEWYN PEART

Nicola de Wit graduated from the University of Otago with a Bachelor of Laws (Hons) and Bachelor of Science in 2011. In 2012 she was admitted to the bar and joined EDS to work as an in-house lawyer. She worked on the Ruataniwha Dam case and the Chatham Rock Phosphate and Trans-Tasman Resources deep-sea mining cases. She also appeared as a junior in the Supreme Court in the King Salmon case which she found 'an awesome experience'.

De Wit can recall being initially thrown in the deep end. 'I remember turning up to mediations and telling myself, "okay, just do it". It was definitely a great experience for a young lawyer to have all these amazing opportunities to give things a go. And it's amazing what you can achieve when you do. I also enjoyed working with some really great legal minds and people in the industry.' As she reflected, 'EDS offers a voice that otherwise isn't necessarily in the system.'

In 2015 de Wit left EDS to work for Chapman Tripp and she is currently a senior associate in the firm, specialising in environmental, planning and resource management law.

grant consent to intensive farming activities for periods up to 20 years, which it duly proceeded to do. The council was, in effect, guaranteeing that applications would be approved. By October 2016, 160 consents for intensive farming and dairying conversions had been granted. None had been declined.

The rationale for this approach was based on changes in the computer software tool called OVERSEER which estimated annual losses of nitrogen and phosphorus from farm systems. When applied to the farms in the Manawatū-Whanganui region, a new version of OVERSEER had produced much higher figures for the leaching of nitrogen. This meant that instead of about 80 per cent of farmers qualifying for controlled activity status (which meant they required consent but the council must grant it), this figure fell to just 20 per cent. As a result, most intensive farmers would require a restricted discretionary consent to farm, which could be declined.

When EDS first heard about the council's resolution and consenting practices, it knew it must be unlawful. The society decided to hold the council to account. Lawyer John Burns and EDS in-house lawyer Madeleine Wright worked on the case for EDS in close association with Fish and Game, which provided funding assistance.

First, they tried to persuade the council to change its approach, but quickly hit a brick wall. As Wright recalls, 'our aim was to work with the council, and keep positive relationships with them, so we could find a way forward together. But the council was very hostile. We were telling them they were acting contrary to council obligations under the law, and the legal process of consenting

The Horizons Regional Council granted 160 consents for intensive farming and dairying conversions for periods of up to 20 years in contravention of its regional plan. The Environment Court found that the potential environmental impact of the consented activities was 'very significant'. Shown here is the Manawatū River running through an intensively farmed area. LLOYD HOMER, GNS SCIENCE

and administering the plan, which were pretty strong accusations.'

The 'softly, softly' approach was failing to work, so the lawyers set about building a case to go to court. As Wright recalls,

> the research for the case was a massive effort. There was 18 months of back and forth and in-depth research before we were ready to lodge the declaration proceedings. We were trying to get information from the council as to what was happening with its water bodies, where consenting was at, the way they were approaching consents and the number of people without consents who were going to require them. Some of the information we were given or told was not quite accurate. We probably had six face-to-face meetings where we were told that the nitrogen levels were decreasing. We said 'are you sure?'.
>
> In the end we lodged an information request under the Local Government Official Information and Meetings Act [which has similar provisions to the Official Information Act but applies to local government]. Initially we were going to get an email back saying it would cost some enormous amount to give

The Manawatū River. The Horizons Regional Council was determined to grant consent for intensive farming activities in contravention of its regional plan. EDS was equally determined to hold the council to account in the Environment Court on water-quality issues. ROB SUISTED/ NATURESPIC

us the information. I responded that, under the RMA, they should have all the information available. In the end we had to send someone in to get it.

We asked freshwater scientist Kate McArthur to help interpret some of the material we were given and commissioned Helen Marr to do a planning assessment. There had been 100 or so consents issued and we selected a handful for Helen to go through in detail. She went through them and pointed out all the deficiencies and inaccuracies.

Proceedings were finally lodged with the Environment Court in September 2016. EDS sought a declaration that the council had failed to correctly apply the statutory requirements and provisions of the One Plan. At the hearing, held in February 2017, Helen Marr gave

The Whanganui River is another water body that is under the management of Horizons Regional Council. It is shown here heavily laden with silt. LLOYD HOMER, GNS SCIENCE

evidence that none of the applications she had reviewed included an assessment of the proposed activity against any of the relevant objectives and policies in the One Plan. She also considered the reasons given in the council's decisions to be inadequate.

It was a compelling case. In its decision released in March 2017, the court found that the 'potential environmental impact of the activities in question is very significant', that over a number of years the council had been unwilling to work with EDS and Fish and Game to remedy the situation, and that 'A public statement from the Court that such an attitude on the part of a law-making and law-administering body is not acceptable is more than justified.'

In noting that the council had been guaranteeing and granting resource consents for periods extending up to 20 years for existing farming activities, the Court observed that this called 'into question whether there has been responsible exercise of the Council's resource management functions'. It was very strong criticism of council performance coming from the courts.

As Madeleine Wright later observed, 'the case was a big wake-up call for councils that they can be held to account'. It was hard to believe that, in this day and age, a council could so blatantly act in contravention of the law to the detriment of the environment. But it was also hard to believe that the country's oversight bodies, the Ministry for the Environment and Department of Conservation, stood by and let it happen. Once again, the matter was left to the efforts of a small group of under-resourced environmental advocates supported by dedicated lawyers. As EDS director Barry Barton observed, 'often our work has been to hold councils' feet to the fire, which is an important role'.

14. Better managing the sea

THE BULK OF EDS'S LITIGATION has been to do with the coast, landscapes and waterways. But most of New Zealand, around 5.8 million square kilometres, is in fact sea. The legal framework for the management of the country's oceans is complex, and much of it is antiquated. This means that legal avenues for challenge are not so clear-cut. Much of EDS's impetus has therefore been to modernise and strengthen the law. However, this has not been easy.

A KEY ELEMENT of the way New Zealand's oceans are managed is through the quota management system (QMS). This cap-and-trade market-based mechanism was brought in during the mid-1980s to address plummeting fish stocks and too many fishing boats chasing too few fish. The QMS was introduced during the Fourth Labour Government, which also introduced a raft of neoliberal policies featuring deregulation and market-led restructuring.

The QMS was based on the idea that over-fishing was caused by fish being available to all, as a public commons, and that this prompted fishers to harvest as many fish as possible, as quickly as possible, before other fishers caught them. This effectively drove a race to get the 'last fish', that is, to exploit the resource before it disappeared. Economists argued that the solution was to create private property rights in fisheries in the form of tradeable quota shares. This, they contended, would incentivise good management of fish stocks through giving the fishing industry a vested interest in the ongoing health of the fisheries.

Although originally granted to the people who actually caught the fish, much of the quota was eventually bought up by large companies and is now effectively controlled by a few powerful corporate interests. Quota was also used to settle Māori Treaty of Waitangi claims over the country's fisheries, and so another sizeable portion is owned by a collective of Māori iwi entities. All this has served to create powerful lobby groups that have opposed environmental restrictions on commercial fishing activity.

Previous pages: Waitawa, Tōtaranui Queen Charlotte Sound. RAEWYN PEART

—000—

EDS'S FIRST FORAY into the marine space after its re-establishment was prompted by the government's efforts to develop New Zealand's first oceans policy. These efforts followed a 1999 report by the Parliamentary Commissioner for the Environment, Dr Morgan Williams, that chronicled widespread dissatisfaction with the current marine management regime. Particular concerns included the unsustainability of fishing, lack of marine protection, poor implementation of the Treaty of Waitangi and a paucity of marine science. Williams recommended that the government establish a task force to develop a national oceans strategy.

In response, Prime Minister Helen Clark tasked Pete Hodgson, who was then Minister of Fisheries, as well as Energy, and Research, Science and Technology, to develop an oceans policy for New Zealand. The process set off with some enthusiasm, but before it was able to deliver, it ran up against an impenetrable roadblock: Māori rights and interests in the marine area had not been resolved.

Prompted by the expansion of aquaculture in their rohe, and being unsuccessful in obtaining aquaculture consents, eight iwi from the top of the South Island, Te Tauihu o Te Waka a Māui, successfully argued that they had unextinguished customary rights over the foreshore and seabed in the Marlborough Sounds. In June 2003, the Court of Appeal unanimously confirmed that the Māori Land Court had jurisdiction to determine whether or not areas of foreshore and seabed were Māori customary land.

The decision created the spectre of the public being excluded from large areas of the coast which might effectively become privately

The quota management system (QMS) was brought in during the mid-1980s to address plummeting fish stocks and too many fishing boats chasing too few fish. Shown here is the inshore commercial fishing fleet in Whakatū Nelson. RAEWYN PEART

owned Māori land. Prime Minister Helen Clark was quick to react, and within days she promised to legislate to overturn the court decision. The announcement, and subsequent detailed proposals that emerged, proved highly controversial.

The Waitangi Tribunal held an inquiry into the government's policy in early 2004 and found it to be in breach of the Treaty of Waitangi. The Human Rights Commission found the policy to be discriminatory. When the Foreshore and Seabed Act was passed on 24 November 2004, Labour Māori MP and junior minister Tariana Turia resigned from parliament. She then formed the Māori Party to contest the subsequent by-election in the Māori electorate of Te Tai Hauāuru.

The uproar had a chilling effect on any policy to do with the oceans. The Oceans Policy Secretariat, a multi-departmental team which had been established to develop the oceans policy, was disbanded. Wanting to keep the prospect of an oceans policy alive, EDS decided to undertake an investigation into the topic, and the society looked to similar efforts in Australia, Canada and the United States to provide inspiration. Conservation International agreed to provide financial support for the project.

The resulting report, *Looking Out to Sea: New Zealand as a Model for Ocean Governance,* was published in 2005. It concluded that

The expansion of mussel farms in the Marlborough Sounds, shown here at Forsyth Bay, prompted eight iwi from Te Tauihu o Te Waka a Māui (the top of the South Island) to claim customary rights over the foreshore and seabed. This led to the foreshore and seabed controversy which had a chilling effect on marine policy during the 2000s. RAEWYN PEART

governance of the country's oceans was not up to the task of providing protection for marine biodiversity, managing the environmental effects of fishing, reducing sedimentation and pollution of the marine area, preventing the incursion of damaging invasive species or managing the environmental impacts of activities outside territorial waters.

EDS proposed that New Zealand, as a small country with a unicameral parliamentary system, was in a unique position to be an exemplar of best-practice oceans governance for the rest of the world. However, it was not to be. The political complexities of unresolved interests in the marine area effectively stymied progress at a national level and this is still an issue today.

—ooo—

IT WAS NOT LONG after the oceans policy project was completed, that the dire state of New Zealand's smallest dolphins came to EDS's attention. As Raewyn Peart recalls,

> in around 2008 or 2009, EDS was contacted by marine scientist Dr Liz Slooten, who was concerned about litigation initiated by commercial fishing interests seeking to overturn protections for Hector's and Māui dolphins. She feared that the Ministry of Fisheries would not have enough resolve to win the case.
>
> Liz was based in Dunedin, but must have travelled to Auckland, and I remember meeting her in my living room in Point Chevalier to get a briefing on the issue. What I heard was a wake-up call. Foremost, although I had lived in New Zealand most of my life and had been working on marine issues for some years, I was unaware that New Zealand had its own endemic species of small coastal dolphin. In fact, my knowledge of dolphins was pretty much confined to what I had seen watching Flipper on TV when I was a kid. When I heard that these dolphins were being killed in fishing nets I was horrified.

Professor Liz Slooten is shown here photographing Hector's dolphins in Akaroa Harbour. With her partner, Professor Steve Dawson, Slooten has long been an advocate for better protection for Aotearoa New Zealand's dolphins. RAEWYN PEART

EDS decided to take up the cause. However, not being a submitter in the first instance, the society had to seek leave to join the High Court proceedings, and this was not approved by the court. In any event, the commercial fishers' legal challenge turned out to be largely unsuccessful in relation to the Hector's dolphin, but it did delay measures to protect the critically endangered Māui dolphin off the North Island's west coast.

As a result of this experience, EDS decided to investigate the adequacy of the legal and policy framework to protect New Zealand's marine mammals. If the country's small endemic dolphins were being killed in fishing nets, clearly something was not right. EDS was able to obtain funding for the project from the New Zealand Law Foundation, an entity which focused on supporting legal research and education. EDS lawyer and policy analyst Kate Mulcahy (now Storer) led the project.

As part of the investigation, Raewyn Peart travelled to Muaupoko Otago Peninsula to meet up with sea lion researcher Amélie Auger. Auger was undertaking research on female sea lions, which were breeding at Sandfly Bay on the peninsula. There were concerns about the impact of the deep-sea squid fishery on the population of endemic New Zealand sea lions breeding in the Auckland Islands, with animals feeding on squid being caught in trawl nets. Otago Peninsula was the only place on the New Zealand mainland where the sea lions were now breeding and it provided an accessible location to learn more about their lives and, in turn, help ensure their survival.

In the resulting publication, *Wonders of the Sea: The Protection of New Zealand's Marine Mammals*, released in 2012, EDS concluded that the Marine Mammals Protection Act urgently needed a comprehensive review. This has still to be undertaken.

—ooo—

IN LATE JANUARY 2012, members of the public reported a dead Bryde's whale near Waiheke Island in Tīkapa Moana Hauraki Gulf.

Bryde's whale.
ROCHELLE CONSTANTINE, UNIVERSITY OF AUCKLAND – WAIPAPA TAUMATA RAU

Rangers from the Department of Conservation subsequently found the carcass floating north-east of Tiritiri Matangi and towed it to Motuihe Island. A necropsy of the 15-metre long female revealed that it had been alive when it received a death blow, almost certainly from a ship. The dead whale was blessed by representatives of Ngāti Paoa and buried on the island.

The incident followed on from an earlier Bryde's whale death just five months previously. In fact, there had been 41 known Bryde's whale deaths in the Hauraki Gulf over the past 16 years. Scientists pinned the blame for many of these deaths on ships transiting the gulf.

Up to 50 Bryde's whales are regular users of the gulf at any one time, out of a New Zealand population of around 150. The whales hunt small fish and crustaceans, which they strain through enormous baleen plates in their mouths, after taking large gulps of seawater. Bryde's whales are long and sleek, growing up to 15 metres in length, and they have a characteristic small, curved dorsal fin located far down their backs.

Although present in other countries, the Bryde's whales are considered to be critically threatened in New Zealand. Part of the reason for this is that the place where Bryde's whales spend much of their time is in an area of the Hauraki Gulf where ships come and go from the country's busiest port at Tāmaki Makaurau Auckland. The whales are long-lived and slow breeders, so the average of two ship-strike deaths a year was not sustainable for the small population.

On reading about the whale death in the media, EDS members became concerned. Gary Taylor and Raewyn Peart discussed the matter with Tim Higham, the executive officer from the Hauraki Gulf Forum,* and Dr Rochelle Constantine, a marine mammal expert from the University of Auckland.

* A statutory body, established under the Hauraki Gulf Marine Park Act 2000, which is tasked with promoting integrated management and the protection and enhancement of the Hauraki Gulf.

Kate Storer (née Mulcahy) worked for EDS as a legal researcher and policy analyst from January 2008 to January 2010 and then again from October 2011 to July 2013. Prior to joining EDS, she completed a Bachelor of Laws (Hons) from the University of Bristol and a first class Masters degree in Public International Law from the University of Auckland. Storer is now a partner in law firm Berry Simons, specialising in environmental law. She continued her association with EDS through being a director from 2013 to 2022.

RAEWYN PEART

In reflecting back on her time at EDS, Storer recalls, 'at that stage I was working from home and EDS just had a small room in a Grey Lynn villa where I used to go every week or so. It was pretty awesome, having just left university, to embark on writing some huge works on these topics. I remember going down to Fiordland and doing a trip out on Doubtful Sound for EDS's dolphin research. That was really cool.

'What was important to me at the time was that even though EDS hired people straight out of university or with only a few years' experience, there was a willingness to put faith in us. EDS took the position that we had good brains that could be put to use. It was really quite special and not something you come across very often, especially in the legal world.'

There had been no action from government agencies to address the problem. From EDS's perspective, they were sitting on their hands while whales were dying. The Department of Conservation had the prime responsibility to look after protected species, including marine mammals, but it was undergoing a major restructure which had diverted resources and staff attention. Other agencies had adopted a policy of deferring to the department on such matters.

Given the urgency of the issue, lack of government action and complexities of regulating the international shipping industry, it seemed worth trying a collaborative approach to see if the shipping companies could be persuaded to adopt voluntary measures to protect the whales. The success of the Land and Water Forum provided inspiration.

The three organisations teamed up to convene a workshop of key stakeholders in early March 2012. It was an invitation-only event in order to create a safe place for the industry to participate. Invitees included iwi, shipping lines, ferry- and boat-based tourism companies, the Royal New Zealand Navy, the Ports of Auckland, Department of Conservation, Auckland Council and environmental NGOs.

At the workshop, Rochelle Constantine and her colleague, scientist Aguilar Soto, summarised the science to date, and advised that international experience showed that reducing vessel speed to 10 knots would significantly reduce the deadly impacts of ship strike on the whale population. At the lower speeds, the whales would have a chance to swim away, and even if they were hit, the injuries were less likely to be lethal. At that stage, large vessels were regularly transiting the Hauraki Gulf at speeds close to 15 knots.

EDS lawyer Kate Storer had investigated possible legal mechanisms which could be used to protect Bryde's whales from ship strike and this was presented at the workshop. She identified rules in the Auckland regional coastal plan as a possible response, along with maritime rules under the Maritime Transport Act, and the establishment of a marine mammal sanctuary under the Marine Mammals Protection Act.

This served to demonstrate that regulation could be used to address the issue and was designed to incentivise industry to commit to the voluntary collaborative process. The industry knew that EDS was not slow to go to court, if necessary, to resolve an issue.

A second workshop soon followed. At this event, the shipping industry made it clear that they were reluctant to slow down ships due to cost and operational impacts. To meet their schedules, the vessels might be forced to travel faster for the rest of the journey, which could incur additional fuel costs.

So the industry looked for other measures to mitigate the higher speeds they wished to maintain, such as stationing extra lookouts, reporting when whales were sighted, and changing the designated route into Auckland so ships travelled over a smaller area of the gulf. They also looked to do additional research into potential technologies that could address the problem, such as using alarms or GPS trackers on the whales and electronic buoys.

But as the scientists undertook more research into the issue, it was clear that slowing down ships would be the only effective option to prevent lethal whale strikes. Further analysis confirmed that vessel activity funnelling into the Hauraki Gulf overlapped with whale movements and there was no obvious area that vessels could use to avoid the whales. This meant that re-routing vessels, which was a mechanism that had been adopted overseas where whale strike issues arose, was not a viable mitigation option in the gulf.

By the time a third workshop was held in early March 2013, there had been a further whale death. This served to highlight the urgency of the issue. The International Fund for Animal Welfare offered to fund researchers in the United States to analyse AIS data (an automatic vessel tracking system) for ships transiting the gulf which could track their speed. There was also discussion of making a speed reduction proposal to the International Maritime Organization which oversaw international shipping.

Later that year Ports of Auckland issued a 'Hauraki Gulf Transit Protocol for Commercial Shipping'. The protocol did not require vessels to keep below 10 knots but advised that 'There is good scientific evidence that the risk to whales is substantially less from ships travelling at 10 knots compared to 15 knots or more' and 'when planning your route to and from Auckland, where possible allow for your vessel to reduce speed when transiting the Hauraki Gulf'. The port company helpfully circulated the protocol to shipping lines and ships' masters.

A fourth workshop was held at the end of 2013. It was now more than 18 months since the collaborative process had begun. There was a lot of activity going on, including research and the like, but there was no sign that the ships were going to voluntarily slow down. EDS and the other sponsoring organisations were becoming frustrated. They invited Ian Lancaster from Maritime New Zealand to give a presentation on options for international recognition of a voluntary protocol at the International Maritime Organization which would need to be proposed by the New Zealand government.

Although the protocol to slow down ship speeds in the Hauraki Gulf would be voluntary, the shipping industry knew that if it was approved by the international authority, it would appear on electronic nautical charts and ships' captains would abide by it as a matter of course. It was something they were keen to avoid.

By September 2014, another dead whale had turned up, and the necropsy confirmed it had been killed by a ship. The AIS analysis showed that 88 per cent of the nautical miles travelled through the Hauraki Gulf by ships were at speeds of more than 10 knots. But the industry was still intransigent on ship speed.

A further series of workshops was convened by Tim Higham from the Hauraki Gulf Forum. It was time to wield the 'big stick'. The option of notifying the voluntary protocol with the International Maritime Organization was again discussed, as was the prospect of briefing the ministers of transport, conservation and environment on the proposal. Ian Lancaster prepared draft papers for consideration, making the proposal even more tangible. Such a move was still strongly opposed by the industry and meetings were tense. But the ongoing discussion of international measures, drafting of actual documents, and discussion of the potential involvement of ministers meant that the internationally notified protocol became a real threat to the industry.

Eventually, these efforts paid off and there was a notable change in attitude by shipping interests. By 2016, data showed that the average vessel speed had dropped to 10.7 knots, compared to 14.2 prior to the collaborative process commencing. Although this was a reduction of only a few knots, it made all the difference. There had been no new reported whale deaths since the September 2014 incident. Rather than being a contestation of views, the workshop became a celebration of achievements.

There have been no further recorded whale deaths from ship strike in the Hauraki Gulf. The strategy of collaborating with the industry had taken time, but had paid good dividends in the end.

AFTER BECOMING AWARE of the plight of dolphins in New Zealand, Raewyn Peart was keen to inform the wider public about the wonderful dolphins around the country's shores and the need to better look after them. This led to her researching the history of New Zealanders' relationship with dolphins and current threats to the various species. She investigated the early dolphins that had interacted with humans — Pelorus Jack and Opo, the history of captive dolphins in marinelands around the country, and how scientists and others sought, often in very trying conditions, to better understand these amazing creatures and address threats to them.

The resulting book, *Dolphins of Aoteaora: Living with New Zealand Dolphins* was published by Craig Potton Publishing and shortlisted for the 2015 Royal Society of New Zealand Science Book Prize. The book was dedicated to the scientists who studied dolphins, and through whose work the plight of dolphins had been brought to the broader public consciousness. Close engagement with

In order to inform the wider public about the wonderful dolphins around the country's shores (including bottlenose dolphins shown here) and the need to better look after them, EDS policy director Raewyn Peart wrote a book on the history of New Zealanders' relationships with dolphins and the current threats to the various species. TIIA COOKE/ KAHURANGI PROJECT

scientists continues to be a hallmark of much of EDS's work.

Protecting New Zealand's extraordinary marine mammals is still a work in progress. Although greater protections have been put in place for the Hector's and Māui dolphins, and the ship-strike threat to Bryde's whales in the Hauraki Gulf has been headed off, the Marine Mammals Protection Act has yet to be reviewed and modernised and many dolphins are still killed in fishing nets each year.

—000—

DURING THE OCEAN POLICY process, it became clear that there was a gaping hole in ocean law, with no legislation to manage the environmental impacts of activities within the exclusive economic zone (EEZ), which covers around 4 million square kilometres of ocean and is around 15 times the size of the country's land area. At the same time, there was increasing interest in mining New Zealand's deep-sea mineral resources. Being located in an active volcanic zone, the country's seabed has a rich variety of minerals, including phosphate nodules; manganese nodules containing cobalt, nickel and copper; polymetallic sulphides rich in copper, zinc, lead, gold and silver; and methane gas hydrates. In 2010, EDS decided to undertake a piece of policy work to investigate the design of legislation to fill this gap in environmental law. The work was led by Raewyn Peart assisted by EDS lawyers Kelsey Serjeant and Kate Storer.

It was around this time that the BP Deepwater

During 2010, EDS initiated a piece of work to investigate the design of environmental legislation for the EEZ. The work was carried out by EDS policy director Raewyn Peart, assisted by EDS lawyers Kelsey Serjeant and Kate Storer. Shown here is Serjeant (second from left) and Storer (in the foreground) on an EDS marine field trip. Also in frame are coastal planner Tom Fitzgerald (left), former EDS in-house lawyer Natasha Garvan and environmental writer Lucy Brake (right). RAEWYN PEART

Horizon oil spill occurred in the Gulf of Mexico. It was the largest marine oil spill in the history of the petroleum industry. Eleven workers were killed and 17 others injured. Around 4.9 million barrels of crude oil were released into the ocean, causing extensive damage to marine and wildlife habitats as well as the gulf's fishing and tourism industries. The incident graphically highlighted what could go wrong if the regulatory framework was lax.

EDS closely examined the regulatory framework that applied to the petroleum industry in the United States, as well as approaches taken in Canada, Australia and the United Kingdom. This highlighted the paucity of controls in New Zealand. All the other countries had specific management bodies for their EEZ, undertook marine spatial planning for the area, and had legislation to create marine protected areas within it.

New Zealand lacked all of these features. The resultant EDS report titled *Governing our Oceans: Environmental Reform for the Exclusive Economic Zone* recommended that either the Resource Management Act (RMA) be expanded to apply to the EEZ, or new environmental effects legislation be promulgated for it. In the end, the government pursued the second option, and the Exclusive Economic Zone and Continental Shelf (Environmental Effects) Act was passed in September 2012.

EDS made extensive submissions on the Bill seeking to strengthen its environmental protections, and the society was mentioned by MPs from several parties as it was debated in the House. When introducing the Bill to Parliament, National Party Environment Minister Nick Smith stated, 'I acknowledge the support of groups like the Environmental Defence Society and the Petroleum Exploration and Production Association, which have welcomed this move as they know it will provide better environmental protection as well as certainty of process ...'.

Later in the debate, Labour MP Charles Chauvel responded,

> The legislation itself is deficient ... It would be much better ... to look at the sort of exercise that the Resource Management Act requires to be carried out when assessing any sort of activity. This is what the Environmental Defence Society asked the Minister to do. He should be very careful before he claims the society's support, because the Resource Management Act calls for the promotion of the sustainable management of natural and physical resources.

David Clendon from the Green Party also specifically referred to the EDS policy paper during the debate stating,

> We are fortunate — and I am sure the Minister will acknowledge that we are fortunate — to have a very substantive independent paper prepared by the Environment Defence Society and published earlier this year under the title Governing our Oceans. We wholeheartedly endorse a number of recommendations from the paper. Not the least of those is the call for a royal commission of inquiry to undertake a comprehensive review of all of the existing mechanisms of international best practice and the current legislative framework, including the territorial sea, the exclusive economic zone, and the continental shelf ... We entirely support the necessity of giving a level of protection to the exclusive economic zone sooner than later, as the Environmental Defence Society has indicated.

Although the Act turned out to be weaker in terms of environmental protection than EDS would have liked, it proved its mettle when two deep-sea mining applications were considered. The first was to mine phosphate nodules off the Chatham Rise and the second to mine ironsands off the Taranaki coast. EDS was involved in both cases, alongside many others, and the cases were declined on environmental grounds by the Environmental Protection Authority.

The government subsequently stripped the authority of decision-making powers for future such consents, instead giving them to government-appointed boards of inquiry. It was rumoured that this was in response to the applications being declined. EDS still considers that the RMA and legislation on the EEZ should be combined rather than having two different statutory regimes applying to different parts of the marine area.

Reflecting back on the process leading up to the passing of the legislation, former Environment Minister Nick Smith commented,

> EDS has been the lead non-government thinker and prodder on trying to get a more comprehensive and systematic approach to the ocean environment. It resulted in the legislation I sponsored through parliament for the EEZ. It, for the first time, required a proper environmental assessment of new activities proposed in that massive area 20 times New Zealand's landmass. I would go so far as to say that I don't think that reform would have occurred without the advocacy and thoughtfulness of EDS.

COMMERCIAL-SCALE aquaculture first became established in New Zealand during the

There was a 'gold rush' for marine farming space in the Marlborough Sounds during the 1980s and 1990s. Shown here is a mussel-harvesting barge in Te Hoiere Pelorus Sound.
RAEWYN PEART

late 1960s, at a time when there was limited environmental awareness. Applications only had to meet minimal requirements and the marine space was effectively 'free', apart from the modest application costs. A 'gold rush' for marine farming space developed during the 1980s and 1990s with councils being flooded with applications. The Marlborough Sounds, in particular, became festooned with hundreds of mussel farms.

To bring things back under control, in 2002 the government placed a moratorium on the further granting of marine farming consents. This was lifted three years later, after more restrictive legal provisions had been put in place. These required marine farms to be located within aquaculture management areas to be identified in regional coastal plans, with marine farming prohibited outside these areas.

But, for a variety of reasons, councils were slow to make spatial provision for marine farming and growth of the industry slowed. So, around the same time as consideration was being given to the management of the EEZ, the government decided to loosen the provisions applying to aquaculture. This was aimed at 'kickstarting' the aquaculture industry to reach its slated $1 billion potential by 2025. The Aquaculture Legislation Amendment Bill (No. 3) was introduced into the House in November 2010.

EDS made extensive submissions on the Bill, which it considered to be a retrograde step. It removed the requirement for marine farms to be established within aquaculture management areas, potentially reinstating a free-for-all. It also gave the Minister of Aquaculture power to directly insert provisions into regional coastal plans over the top of councils. EDS thought this was ministerial over-reach, enabled arbitrary rule-making and set a dangerous precedent.

While the amendment Act went through the parliamentary process, various MPs picked up on EDS's submissions. Green Party MP Gareth

EDS undertook an examination of the regulatory regime for marine farming in New Zealand in 2019 and concluded that spatial marine management needed to be strengthened. Shown here is a mussel farm off the coast of Te Tara-o-te-Ika Coromandel Peninsula. RAEWYN PEART

Hughes observed during its first reading,

> According to the Environmental Defence Society and other experts, best practice involves heading towards integrated marine spatial planning. This bill does not create a comprehensive integrated mechanism to replace the regional aquaculture management areas ... we agree with the Environmental Defence Society's concern that the representation of industry interest, as well as responsibility for setting environmental standards, the drafting of national policy and the management of the environmental impacts of aquaculture, are too many functions to be vested in one agency ... In a nutshell, it is not good policy or practice for the industry's promoter also to be the regulator.

In the end, the government passed the amendment act with little change. EDS later undertook a specific examination of the regulatory regime for marine farming, funded by Fisheries New Zealand, and published the results in the 2019 publication *Farming the Sea*. This recommended a strengthening of spatial marine management, a more flexible regulatory regime to enable marine farms to move to more suitable sites, and stronger risk management. It also argued for a more developed allocation and de-allocation framework for marine space.

EDS considered there could be a win-win situation where a new regime could create stronger environmental protections whilst enabling the industry to operate more efficiently.

—ooo—

MARINE PROTECTION has always been a fraught area in New Zealand. As described earlier, EDS had been involved in the early implementation of the Marine Reserves Act 1971, lodging the first marine reserve application for the Poor Knights Islands. The legislation was ground-breaking at its time, being possibly the first legislation in the world to provide for the spatial protection of the marine environment.

But by the 2000s, expansion of the marine reserve network had stalled and it was clear that the legislation needed to be updated. Foremost, it did not mention biodiversity protection, with the ostensible purpose of marine protection being only for scientific study. In addition, there was no provision to create marine reserves within the EEZ.

By 2002, a Marine Reserves Bill had been introduced into Parliament by Conservation Minister Chris Carter and it was referred to a select committee for consideration. But that's as far as it got. It languished there for years due to lack of political support. The pro-fishing fraternity, including commercial and recreational fishers, was powerful enough to halt the new law in its tracks.

Meanwhile, struggles went on to protect areas around the coast under the old legislation. One of these played out at Te Pātaka-o-Rākaihautū Banks Peninsula. The proposal to establish a small marine reserve near Dan Rogers Bluff at the entrance to Akaroa Harbour was marked by years of controversy. The 530-hectare reserve was first formally proposed by the Akaroa Harbour Marine Protection Society in December 1995, when it lodged a formal application with the Department of Conservation. The area contained spectacular volcanic cliffs, sea caves and sea stacks.

A Department of Conservation report described the proposed marine reserve area as follows:

> The underwater topography consists of cliffs and bluffs falling vertically to the seabed and colonised by a rich diversity of plant and animal communities. At the base of Dan Rogers Bluff there are huge room-sized boulders that provide spectacular underwater scenery and habitat for different types of marine communities that are typical of parts of the exposed Banks Peninsula coastal environment.
>
> Sub-tidal communities comprise a colourful mosaic of species including giant bull kelp beds, green mussels, sea tulips, hydroids, sponges, sea squirts and sea anemones.
>
> The seabed supports populations of horse mussels, encrusting sponges, sea squirts, red algae, cushion, snake and biscuit stars; and a variety of tubeworms, molluscs and bivalves. The reef around Gateway Point supports an extremely rich and diverse fauna and flora, with at least 10% of the benthic species found in this area being 'undescribed'.
>
> The area also supports a diverse array of southern New Zealand fish species, including both rocky-reef fish and inshore pelagic fish, some of which have commercial value.

Notwithstanding the richness of the habitat to be protected, and the small size of the proposed reserve (covering just 12 per cent of Akaroa Harbour), the proposal was strongly opposed by recreational fishers who would no longer be able to fish in the area. In addition, the local rūnanga was concerned about the loss of customary fishing rights.

When the application was publicly notified in January 1996, a total of 3043 submissions were received; 709 objections and 2334 in support. The processing of the application was then effectively halted when recreational fishers proposed an alternative (and much smaller) marine reserve just outside the harbour in Pōhatu Flea Bay

The proposal to establish a marine reserve near the entrance to Akaroa Harbour proved highly controversial. Seen here is a kayaker in the harbour protesting in favour of the marine reserve. RAEWYN PEART

in January 1997. In addition, Te Rūnanga o Koukourārata, Ōnuku Rūnanga and Wairewa Rūnanga proposed a taiāpure (a Māori spatial management tool for fisheries) inside Akaroa Harbour. These were both subsequently created, but they excluded the initial area proposed for the Akaroa marine reserve, thereby leaving open the possibility that it could proceed.

Once the Akaroa taiāpure was created in 2006, the Department of Conservation continued processing the initial marine reserve application. More than a decade had now passed since it had first been lodged. The department undertook a second round of public consultation which generated 77 submissions of which a third (25) were opposed to the reserve.

Then, in August 2010, Conservation Minister Kate Wilkinson declined the application. This was ostensibly on the basis that it would adversely affect existing recreational fishing. Gutted, the Akaroa Harbour Marine Protection Society approached EDS for advice on what to do through the society's community advice service. Natasha Garvan took the phone call. She recalls, 'I remember getting a request from the Akaroa Protection Society. They were asking for advice on judicial review. I looked at the decision, identified what the error of law was, and advised there was a strong case. They then got advice from another lawyer who said it was not a good case. But Rob Makgill and Rob Enright took it on and they won.'

The minister had taken a narrow view of the tests in the legislation, and had only weighed up the impacts on recreational fishing against benefits to recreational use of the area from creating the reserve. She had determined that the latter had not outweighed the former and therefore had declined the proposal. The Akaroa Harbour Marine Protection Society lawyers argued that the minister should have taken a broader view to consider the full merits of the proposed reserve against any adverse effects on recreational fishing.

The High Court was sympathetic to this view. It found that the minister should have been satisfied that the impact on recreation fishing would be 'both excessive and unjustified' in order to decline the proposal on those grounds. It quashed the minister's decision. On reconsideration by a new Minister of Conservation, Dr Nick Smith, a slightly smaller reserve was approved in 2013.

The court proceedings had both enabled the marine reserve to be created and had also clarified the law in a way that was more favourable to future marine reserve

Progress in updating the dated Marine Reserves Act 1971 is still stalled despite multiple efforts to pass new legislation. Shown here is the first marine reserve established in New Zealand (and possibly the world), the Cape Rodney-Okakari Point Marine Reserve (Goat Island) near Leigh. CRAIG POTTON

proposals. As Natasha Garvan reflected, 'It was very satisfying that I gave that advice to the community group, identified the error of law, handed it on to Rob and they took it forward. It was exciting to be able to contribute in that way.'

In 2012, after there had been no progress in passing new marine protected area legislation for over a decade, the government indicated a renewed intention to revise the Marine Reserves Act. In order to influence the proposed new legislation, EDS decided to undertake an investigation into what best practice marine protection now looked like and how it might apply to New Zealand.

The project was financially supported by the Lion Foundation and ASB Community Trusts (now Foundation North). Kate Storer led the work which looked at approaches to marine protection in England, California, the Australian Commonwealth and New South Wales. Abbie Bull also assisted with research. The results were published in *Safeguarding Our Oceans*, which identified three options for the new legislation and provided a series of recommendations for its design.

The report was well received. In the debate in the House on the Subantarctic Islands Marine Reserves Bill in December 2012, then Labour Deputy Leader Grant Robertson stated,

> The Environmental Defence Society has this year put out a publication around its views of what needs to happen in terms of marine reserves legislation. I had the honour of launching it earlier this year. I tend to agree with the society that the time has come not to return to the legislation that Labour had in front of the House 10 years ago, because time has passed that as well, but to create new marine reserves legislation that actually allows for the efficient promulgation of these kinds of reserves, that allows for a creation of a network of different types of marine protected areas, all the way from no-take areas through to the full protected reserves that most people would understand.

Unfortunately, the impasse over legislative reform in the area continued. In January 2016, the National-led Government had another crack at the matter and released a discussion document proposing yet another regime. This, controversially, excluded the EEZ from its ambit and proposed the established of 'recreational fishing parks' in the Hauraki Gulf and Marlborough Sounds which appeared to be more about enabling fishing than protecting the marine environment.

This proposal also stalled. The Labour Government did not come up with its own proposals despite committing to update marine protected areas legislation in its 2020 election manifesto.

Meanwhile EDS keeps pushing for reform.

—ooo—

DURING THE 2000s, EDS's Raewyn Peart had been undertaking consultancy work for the Hauraki Gulf Forum in a private capacity. In 2011, the forum commissioned her to review international approaches to marine spatial planning and to assess the potential application of these approaches to the gulf. This followed on from the forum's 2011 State of the Environment report which had, unusually at the time, used a historical baseline to chronicle the ongoing degradation of marine life and habitats in the Hauraki Gulf. The report garnered considerable attention, with the *New Zealand Herald* headlining 'Hauraki Gulf: toxic paradise'.

Peart's international review, which looked at various marine spatial planning processes in Australia, the United States, Canada, Belgium and Norway, concluded that marine spatial planning was a well-accepted strategic planning process which would help to achieve 'integrated management and the protection and enhancement of the life-supporting capacity of the Gulf'.

The report lay dormant for a while, but as more people read it, gradually the idea of setting up a marine spatial planning process for the Hauraki Gulf took root. The early success of the Land and Water Forum had increased the appetite for using collaborative processes in New Zealand. EDS joined forces with Tim Higham from the Hauraki Gulf Forum to persuade Auckland Council and subsequently Waikato Regional Council to sponsor a collaborative marine spatial planning process for the gulf.

Getting approval for the project, called Sea Change Tai Timu Tai Pari was not easy. It was a year before Auckland Council approved the project, and with councillors still evenly

The Hauraki Gulf marine spatial plan was developed by a diverse collaborative stakeholder group including iwi representatives and commercial fishers, recreational fishers, marine farmers, dairy farmers, infrastructure providers and environmentalists. Members of the group are shown here on a field trip to a dairy farm on the Coromandel Peninsula.
RAEWYN PEART

divided on the matter, it only proceeded on a casting vote by the chair. There were also long negotiations with iwi. In the end the project was overseen by a co-governance entity which had 50 per cent agency representation and 50 per cent iwi.

The marine spatial plan itself was to be developed by a collaborative stakeholder working group consisting of 14 people — four mana whenua representatives and 10 stakeholders — and included commercial fishers, recreational fishers, marine farmers, dairy farmers, infrastructure providers and environmentalists. EDS policy director Raewyn Peart was one of two environmental representatives on the group.

The stakeholder working group first met in late 2013. What followed was a very intense and difficult process, but also a life-changing one for those involved. As Peart recalls,

> when we first met, everyone sat around the room eyeballing each other. I remember looking across at the fishermen and thinking 'they're the enemy'. I was also suspicious of the Māori representatives as I knew that iwi held a lot of fishing quota.
>
> Although I had been involved in marine issues for many years, I realised that I had never actually spoken to a commercial fisherman. As it turned out, the fishermen on the group became some of my closest friends. I found that they also had a passion and love for the sea, because they had spent all their lives there. They had fantastic knowledge of what was happening out there and were equally concerned about dwindling marine life.
>
> It was hard getting agreement amongst the 14 members of the group, and at times we have some very difficult conversations. Efforts to integrate a Māori world view broke down and the project was paused for a while so it could be reconfigured.
>
> It was very much learning as we went. But by the end we were a tight group of 14 who had each other's backs. What turned out to be the biggest challenge was

The Sea Change Tai Timu Tai Pari collaborative group brought diverse interests together to agree on a future for Tīkapa Moana Hauraki Gulf. Better managing sediment from clear-fell forestry harvesting was one of the issues that the group grappled with. Shown here are members visiting the Whangapoua forest. In the foreground is group member Tame te Rangi and to the far right is group member Conall Buchanan. RAEWYN PEART

> getting councils and the government to implement our recommendations. That is still a work in progress, even though we delivered the plan in 2016.

Although implementation of the marine spatial plan has been slow, the process served to build relationships and raise awareness which has supported a myriad other efforts now underway to restore the gulf. In July 2019, Conservation Minister Eugenie Sage and Fisheries Minister Stuart Nash appointed a ministerial advisory committee to assist the government in developing a response to the Sea Change plan. Peart was one of only two members from the Sea Change Stakeholder Working Group appointed to the committee, which reported in September 2020.

The Hauraki Gulf has remained on the political agenda. Both Labour and the Green Party included restoring the gulf in their policy agenda for the 2020 national election, the only marine place to be specifically mentioned. The government has progressed the fisheries and marine protected area parts of the Hauraki Gulf spatial plan and Peart was recently appointed to the Hauraki Gulf Fisheries Plan Advisory Group, which has been established to progress its fisheries recommendations. In late August 2023, new legislation was introduced into Parliament to finally create the marine protected area network that the Hauraki Gulf sorely needs. A Hauraki Gulf Fisheries Plan was also approved by the Minister of Fisheries in the same month and proposals to ban trawling from most of the Hauraki Gulf circulated for public comment.

—ooo—

WHILE PARTICIPATING in the Sea Change process, Peart was also writing an environmental history of the Hauraki Gulf, designed to communicate to a broader audience the need to take action to restore the area. The Hauraki Gulf Forum came on board as a foundation sponsor for the project. Additional support was obtained from

EDS policy director Raewyn Peart wrote an environmental history of the Hauraki Gulf to support efforts to restore the marine area. She is pictured (left) at the launch of the book in 2016, and (above) with Coromandel MP Scott Simpson at the event. TANYA PEART

Auckland Council, the Tindall Foundation, Ports of Auckland, Watercare Services and Foundation North. The output was envisaged as a high-quality coffee-table book and Peart sourced many historical photographs for the publication. Bateman Books came on board to publish the book, which was titled *The Story of the Hauraki Gulf*.

Former EDS general manager Fiona Driver remembers the launch event for the book in October 2016. It was held at the Royal New Zealand Yacht Squadron clubrooms at Westhaven in Auckland. As Fiona recalls, 'It was a crazy night at the yacht squadron. There were so many people packed in there. I remember Raewyn saying that it was quite noticeable that EDS publications had gone from being very technical to more mainstream, which is what she was trying to do with this one. It was pretty exciting to get that profile.'

The book rapidly sold out and was quickly reprinted for Christmas sales. It received many positive reviews, went into a third print, and is still being sold in bookshops. A revised edition is currently being considered.

AFTER THE SUCCESS of the Sea Change Tai Timu Tai Pari process, EDS was keen to see marine spatial planning rolled out to other areas of the country. It decided to both undertake a lessons learnt study of the Sea Change process and also to have a look at more recent international best practice. Former EDS lawyer Kelsey Serjeant, who by this time was living in Australia, agreed to assist with the second phase of the work, which was supported by the Department of Conservation. The resultant report, *Healthy Seas*, was released in 2019.

EDS is still pushing for marine spatial planning to be adopted more widely as a marine management tool. The society has received some support in government proposals for the development of regional spatial strategies, covering both land and the territorial sea, under the proposed Spatial Planning Act.

In order to support the adoption of marine spatial planning in New Zealand, EDS researchers Raewyn Peart and Kelsey Serjeant travelled overseas to research models applied in other countries. Serjeant is shown here on a research trip to California.
RAEWYN PEART

THE MANAGEMENT OF the country's fisheries, which had been under a quota management system since the mid-1980s, was something EDS had been keen to investigate for some time. This was because of the evident issues of localised depletion of fish stocks and habitat damage from fishing methods, such as trawling and dredging, which were not being addressed.

However, the society struggled for years to find a funder for the work. Finally, in 2017, EDS secured funding for the project through the Department of Conservation and Ministry for the Environment's community funds. The money supported case studies of fisheries management in the Hauraki Gulf, Kaipara Harbour and Te Tauihu o Te Waka a Māui (top of the South Island) marine areas.

EDS policy director Raewyn Peart led the project. Early on it became clear that many fishermen, who were out on the water every day as part of their jobs, were extremely concerned about the way fisheries were managed. But they were unable to speak out publicly for fear of losing access to annual catch entitlements, which they needed to cover their catch, and had to obtain from quota owners. So Peart undertook a series of highly confidential interviews with commercial fishermen and sought to tell their stories in the report.

There had also been several rounds of restructuring within the Ministry of Fisheries and Peart found that many former fisheries management employees were also willing to talk under strict confidentiality. In contrast, the Ministry for Primary Industries, which was in charge of fisheries management at that time, failed to respond to requests for interviews.

To see what commercial fishing involved first-hand, Peart arranged to go out on

After completing a conjoint Bachelor of Science (Biology) and Bachelor of Laws degree at the University of Auckland, **Kelsey Serjeant** joined Russell McVeagh's resource management team for two years. During this time, she became aware of EDS's involvement in a number of high-profile Environment Court cases and contacted EDS looking for work. She joined the society in September 2010 as a legal and policy advisor.

Serjeant ran EDS's litigation programme and represented EDS on the Mackenzie Country Forum. She also assisted with policy work. She was co-author of the EDS publication on environmental reform of the EEZ and later, when she worked part-time for EDS remotely from Australia, on marine spatial planning.

COURTESY OF KELSEY SERJEANT

'Coming to EDS felt very much like being thrown in the deep end but at the same time it was exciting. Gary [Taylor] and Raewyn [Peart] trusted me with some key matters right up front. Gary put me in the driving seat for the EDS litigation programme and I was in the Environment Court soon after I arrived. I had the chance to go to mediations on behalf of EDS. For a junior lawyer in their early 20s that was unheard of. I also very much learnt policy work through EDS. I was so grateful to have that experience really early on in my career.'

Serjeant's work on marine spatial planning was a particular highlight. 'Marine spatial planning was in its early days of being a planning practice. I remember doing this amazing trip to California with Raewyn and interviewing some of the most prominent marine spatial planning thinkers, probably globally.

'One of the things that really sticks out as different with EDS is that it has this attitude of "why don't we see if we can get involved in this issue and get a good outcome?" Raewyn and Gary would say "let's see if we can make it work, let's see this as an opportunity", even if they had hundreds of other things on the to-do list that week. No problem was too difficult to solve. In contrast, my experience of leadership in other organisations and government since that time is people saying "what are the problems of getting involved in this? Will it stretch our resources?"'

After leaving EDS in 2012, Serjeant worked for the Parliamentary Commissioner for the Environment for four years, where she led a project on OVERSEER (a model that estimates nutrient leaching from farms) and was a senior project officer for the Department of Environment, Land, Water and Planning in Victoria, Australia.

In order to investigate how well the fisheries management system was operating in New Zealand, EDS policy director Raewyn Peart interviewed many fishermen and went out on commercial fishing vessels to see what was involved first hand, including on this long-liner out from Leigh north of Auckland.
RAEWYN PEART

commercial fishing trips on Danish seining, long-lining and crayfishing boats, As she recalls,

> I was very much a newbie when it came to commercial fishing but was keen to get out on the water to see fishing operations first-hand. I met some very helpful fishermen who agreed to take me out on their boats. I think they were impressed that as an environmentalist I was making an effort to find out what commercial fishing actually entailed.
>
> My trip on the Danish seiner was the most eventful. I had arranged to meet the boat at the Whitianga wharf at dawn. I had to leave during the night to drive down from Auckland, and sure enough, when I arrived in Whitianga bleary-eyed this beautiful vessel was tied up at the wharf. The skipper said 'come on board' and he then headed into town to get provisions. The first thing I noticed when I got on the boat was that there were *Forest and Bird* magazines in the wheelhouse. I found this enormously surprising. I did not expect a commercial fisherman to be a greenie!
>
> Once we started chugging out of the bay I realised that we had not actually confirmed the duration of the trip. I had hoped it was just overnight as I had a meeting in Auckland in two days' time. When I broached the subject with the skipper he just shrugged and said, 'when we head back depends on how well the fishing goes'.
>
> The next thing I noticed was there was no toilet on the boat. I was shown to a bucket on the stern. I needed to make sure I used the bucket between net hauls, otherwise I was in danger of becoming entangled in the enormous wires, ropes and pulleys.
>
> We spent the night off an island, and before dawn chugged out to fish in the Colville Channel. I woke to an absolutely glorious day. The fishermen had been working much of the night packing fish but were up before dawn heading out to the next fishing ground. I learnt that the

One of the most memorable fishing trips EDS policy director Raewyn Peart experienced was on a Danish seiner out of Whitianga, shown here. She was surprised by how hard the fishermen worked, their concern for the ocean and the kindness they showed her. RAEWYN PEART

> skipper had left a crew member ashore so I could have his bunk, so the skipper was up packing fish with the other crewman. It was incredibly hard and repetitive work.
>
> In the end, the skipper made a special trip to drop me back at Whitianga. As I scrambled up on the dock in the fading light he thrust a large package into my hands and then headed back to sea. When I opened the package, I found a whole lot of beautifully filleted snapper.

The resultant report, *Voices from the Sea: Managing New Zealand's Fisheries*, was launched by Fisheries Minister Stuart Nash in April 2018. It was one of the first in-depth analyses of the operation of New Zealand's quota management system. It unpicked the theory behind the system and profiled how its operation in practice was quite different to the commonplace rhetoric that New Zealand had one of the best fisheries management systems in the world. It demonstrated that the country had fallen far behind international best practice and there were serious issues to be addressed. At the end of the report, EDS called for an independent inquiry into the fisheries management system.

The report received a strong backlash from parts of the commercial fishing industry but was well received by many others, including politicians. Te Ohu Kaimoana (the Māori fisheries trust) released a media statement attacking the credibility of the report, although it did not identify any particular flaws with it.

To help defuse the situation, EDS decided to hold a seminar in Auckland to discuss the findings of the report, and invited Dion Tuuta, the then CEO of Te Ohu Kaimoana, to be part of the panel. This resulted in a more positive debate. By this time, the Labour Party had acknowledged the need to reform the fisheries management system.

To assist with his reform efforts, Fisheries Minister Stuart Nash decided to establish a Ministerial Fisheries Advisory Group and proposed to appoint Raewyn Peart to it. However, Peart's appointment was vetoed by Labour's coalition partner New Zealand First.

It was later reported that commercial fishing giant Talley's had made a $10,000 donation to NZ First candidate Shane Jones in 2017 as well as donating $26,950 between 2017 and 2019 to the New Zealand First Foundation.

On 21 November 2018, *Stuff* ran a piece with the headline 'NZ First blocking analyst's fisheries review appointment over bias claims'. A senior fishing industry source was quoted as saying the industry was uneasy about having Peart on the panel as she had published a book that criticised the quota management system.

In the end, Minister Nash decided to abandon the advisory group idea. Although no independent inquiry has yet been undertaken into fisheries management, the government had been tightening the regulation of commercial fishing, including introducing cameras on fishing boats and a no discard rule. EDS is still keen to see stronger environmental manage-ment of fishing, including the phase out of environmentally damaging methods such as trawling and dredging.

MUCH OF EDS'S prior work on marine management has been encapsulated and extended in the society's oceans reform project which commenced in 2020 and is continuing. The project, which was led by EDS policy director Dr Greg Severinsen and has now been picked up by Raewyn Peart, has taken a holistic view of the oceans management system. It has been financially supported by the New Zealand Law Foundation, the Michael and Suzanne Borrin Foundation, the Ministry for the Environment and Department of Conservation.

The need for oceans reform was underscored in a 2021 Cabinet paper setting out the parameters for the government's review of the resource management system. It referred to overlapping marine legislative frameworks being addressed through 'a subsequent review of the marine system'. The Cabinet paper signalled the importance of the EDS work on the government's thinking, stating,

> Civil society and industry organisations are also considering oceans issues, for example, the Environmental Defence Society (EDS) is conducting an oceans research project (partially funded by MfE and DOC) called — Reforming Oceans Management: How to achieve better environmental, social and economic outcomes in our marine environment.... This civil society and industry work may inform the Government's longer-term future work.

The EDS project covers the overarching purpose of an oceans management system and principles that should underpin it, key legislative regimes applying to the marine area as well as institutional settings, funding and management tools. The role of Māori in marine management and the integration of te ao Māori world view into a future system has been a particular focus, with a Māori law academic from the University of Canterbury, Adrienne Paul, joining the project team. EDS also considered international lessons, although Covid-19 prevented an international study tour as well as in-person interviews in New Zealand.

The work explored the potential utility of fundamental reform of the existing legislative and institutional framework and it concluded with four possible approaches to oceans reform. These included building on and improving the current system and more radical redesign and integration, including through a

In 2020 EDS commenced a major project investigating oceans reform and developed options for a new, more integrated management system. The project seeks to more effectively address the multiple and increasing pressures on the marine environment, including sediment from forestry harvesting shown here in Te Whanganui Port Underwood. RAEWYN PEART

new Oceans Act and Oceans Agency. The other two approaches were more exploratory, with one looking at the implications of enlarging the rangatiratanga sphere of Māori governance and the other examining the implications of giving the oceans or parts of it legal personhood, in a similar way to Te Urewera and the Whanganui River, both of which have been given the same rights as people under dedicated statutes, and with guardians appointed to speak on their behalf.

The final report of Phase 1 of the project, titled *The Breaking Wave, Oceans Reform in Aotearoa New Zealand* was launched in June 2022 by then Conservation Minister Kiritapu Allan. EDS is now embarking on Phase 2 of the work, which will develop a single preferred model and chart a careful pathway towards it. EDS will be pushing for oceans reform to be implemented over the coming years.

15. Strengthening the system

AS WELL AS BECOMING involved in specific cases around the country, EDS also saw the necessity of strengthening the resource management system overall. It sought to do this through winning cases in the courts that would set positive precedents. These would strengthen council planning documents, which provide the rules around development, and build the case for change where the law was wanting.

DESPITE EDS OBTAINING a landmark Court of Appeal decision under the Town and Country Planning Act in the Karikari case (see chapter 7), that a 'balancing approach' should not be taken to matters of national importance, once the Resource Management Act (RMA) came into force, councils and the courts reverted back to the old approach. This came to be called the 'overall broad judgement' approach. It effectively watered down any environmental protections as environmental harm could be outweighed by countervailing economic and social benefits.

EDS thought this approach was wrong in law and was keen to challenge it in the higher courts. However, it had to find the right case, as losing could be disastrous and embed the 'overall broad judgement' approach more deeply within the resource management system. Eventually a suitable case emerged in the context of hotly contested salmon farming proposals in the highly scenic Marlborough Sounds.

There was important context leading up to the case. As described earlier, in 2010 the New Zealand Coastal Policy Statement, which had been a high-level and woolly document, was considerably strengthened. The requirement to 'avoid' had been included when referring to adverse effects on outstanding natural landscapes and natural character. At the same time, new provisions had been included requiring councils to provide for marine farming as well as ports in their plans.

The following year, the RMA was amended to provide for 'streamlined provisions' for consenting. Additional provisions were added to the Bill to enable resource consents to be sought concurrently with a request for a plan change. This meant that if a plan did not provide for an activity, changes to the plan could be sought at the same time as consent for the activity, meaning that the individual activity drove the plan change.

The RMA amendments also allowed applications to be lodged directly with the Environmental Protection Authority, a statutory environmental regulator established in 2011 (rather than the council), which would advise on whether they involved matters of national significance. If they did, the application could be 'called in' and heard by a board of inquiry appointed by the Minister of Conservation, thereby circumventing the council's role in determining applications. In addition, applications heard by a board of inquiry could not be appealed to the Environment Court on their merits. Only appeals on matters of law to the High Court were permitted.

The RMA amendments came into force on 3 October 2011. New Zealand King Salmon Limited immediately lodged an application for plan changes and consents to authorise nine new salmon farms at various locations around the Marlborough Sounds. At that time, the company operated six salmon farms in the sounds, so the application represented a considerable scaling up of its operations.

Things moved quickly once the applications were lodged. By 10 October the Environmental Protection Authority had already considered the application and recommended to the Minister of Conservation that it be heard by a board of inquiry. Less than a month later, on 3 November, the minister had appointed a board to determine the matter. It was chaired by retired Environment Court Judge Gordon Whiting.

Previous pages: Tōtaranui Queen Charlotte Sound. The Marlborough Sounds were the focus of an EDS case which went all the way to the Supreme Court and overturned more than 20 years of jurisprudence. RAEWYN PEART

In 2011 New Zealand King Salmon applied for plan changes and consents to authorise nine new salmon farms at various locations around the Marlborough Sounds. This led to litigation which took EDS all the way to the Supreme Court. Shown here is a salmon farm that was eventually established at Waitātā Reach. RAEWYN PEART

The applications were to be considered within the context of the Marlborough District Council's Marlborough Sounds Plan, which had come into effect in 2003. This had already zoned areas for 576 marine farms within the Marlborough Sounds, the vast majority of which were mussel farms, and had excluded aquaculture from other areas through making it a prohibited activity. Eight of the nine sites applied for by New Zealand King Salmon were in areas where the plan prohibited marine farming. It was for this reason that the company sought to change the plan. The Marlborough District Council stood by its planning approach and opposed the applications.

EDS was alerted to issues with marine farming in the Marlborough Sounds by then Green Party MP Steffan Browning. When the EDS team was in Whakatū Nelson to run a community workshop, Browning arranged for a local pilot to fly them over the sounds to see the extent of marine farming for themselves. The day the flight was scheduled was stormy and it was not until late in the day that the cloud lifted briefly and the flight took off. As Raewyn Peart recalled, 'I had heard that there was a lot of marine farming in the Marlborough Sounds, but I was blown away by the extent of it when seen from the air. Bay after bay was filled with long rows of mussel buoys. I could not understand how such extensive areas of the coastline had been given over to the industry, particularly in such an iconic landscape as the Marlborough Sounds.'

EDS decided it needed to get involved.

When the EDS team was in Nelson to run a community workshop, Green Party MP Steffan Browning arranged to fly them over the Marlborough Sounds to view the large number of marine farms there. Seeing the extensive areas of coastline that had been given over to aquaculture, EDS decided to become involved in the issue. Shown here from the left is former EDS in-house lawyer Natasha Garvan, EDS chief executive Gary Taylor and former EDS policy researcher Abbie Bull about to board the flight.

Above right: The EDS flight over the Marlborough Sounds highlighted the numerous mussel lines spread out along the coastline, as shown here. RAEWYN PEART

The society worked closely with local group Save Our Sounds and lodged submissions with the Environmental Protection Authority opposing the salmon farming proposals. In its submissions, EDS was careful to state at the outset that it supported aquaculture in appropriate locations. However, it opposed salmon farms being located in places that impacted outstanding natural landscapes and natural character. To help identify which of the proposed farms had such impacts, EDS sought the help of landscape architect Stephen Brown. As Brown recalled, 'I remember being carted around the islands in a launch with three local councillors on board. One of them owned the launch. They were all desperately worried about the future of the Marlborough Sounds.'

The board of inquiry hearing was held in Blenheim and took 37 days. Presentations were made by 181 witnesses and submitters, 10,400 pages of evidence were submitted and the transcript of the hearing ran to some 4100 pages. The sheer scale of the technical detail threatened to swamp EDS and the other community groups. So, to maximise its impact with the limited resources at hand, EDS focused narrowly on landscape issues, and it only contested two of the sites: the one proposed for Papatua in Te Anamāhanga Port Gore, which impacted an outstanding natural landscape, and the one proposed for Waitātā Reach, which was in a largely unmodified waterway linking Te Hoiere Pelorus Sound to Te Moana-o-Raukawa Cook Strait. Stephen Brown gave evidence for EDS. He had closely examined all the sites and gave detailed evidence about the impacts of allowing salmon farms in the two locations that EDS was opposing.

Port Gore is a large bay located on the very outer edge of the sounds, between Queen Charlotte and Te Hoiere Pelorus sounds. Two long sinuous capes, Cape Lambert and

Te Taonui-a-Kupe Cape Jackson, demarcate each side of the bay. It is one of the largest and least developed bays in the Marlborough Sounds, and a visit there evinces a feeling of wildness and remoteness.

The proposed salmon farm at Papatua was to cover an area of 91 hectares (around 90 rugby fields). This was to be divided into four blocks with up to five circular plastic pen nets 40 metres in diameter within each block. Only two of the four blocks were to be used at any one time with the pen nets being rotated between the blocks. This was because of the low water flow at the site, which meant that there would be poor flushing of nutrients from the fish in the pens. It was the largest of the nine proposed farms.

The board of inquiry released its final decision in February 2013. This approved four of the nine proposed farms including the one at Port Gore. The decision accepted that the Papatua site in Port Gore was an outstanding natural landscape and that 'unquestionably, there will be high to very high adverse visual effects on' the landscape and a high effect on an area of outstanding natural character. The decision also acknowledged that the proposal failed to give effect to policies in the Coastal Policy Statement which required adverse effects on outstanding natural landscapes and areas of outstanding natural character to be avoided. However, the board of inquiry still granted consent to establish a salmon farm on the site.

It did so on the basis that the direction to 'give effect to' the New Zealand Coastal Policy Statement in the RMA did not require every policy to be met. It stated, 'requiring that every single policy must be given full effect to would otherwise set an impossibly high threshold for any type of activity to occur within the coastal marine area'. Instead, the board took the view

When EDS became involved in the salmon farming issue in the Marlborough Sounds, Gary Taylor contacted New Zealand King Salmon Limited and arranged a site visit to better understand what such an activity involved. Shown here from the left is the farm manager, Taylor, former EDS policy researcher Abbie Bull and former EDS in-house lawyer Natasha Garvan. RAEWYN PEART

that it was 'a matter of judgement on the facts of the particular proposal' thereby effectively applying the overall broad judgement approach.

On the facts, the board found that the high landscape and natural character impacts of the farm in Port Gore were outweighed by the ability of a salmon farm in such a remote place to provide a biosecure area away from other salmon farming areas in the Marlborough Sounds. In effect, the biosecurity considerations of a commercial salmon farming operation trumped outstanding public landscape values.

EDS took the decision to the High Court on appeal, alongside Save Our Sounds. Rob Enright ran the case for EDS assisted by in-house lawyer Nicola de Wit. But in August 2013 the High Court dismissed the EDS appeal on the basis that it could not substantially alter the weight given to competing considerations by

Port Gore, one of the largest and least developed bays in the Marlborough Sounds, was the location for a proposed salmon farm covering an area the size of around 90 rugby fields. SHELLIE EVANS

the board of inquiry. This was because, under the new provisions in the RMA, appeals to the court were only available on matters of law and the court considered that the cases argued by the environmental groups went to the merits of the decision which was out of its scope.

At this point Gary Taylor concluded that EDS needed to get the case into the Supreme Court if there was to be any prospect of success. EDS sought leave to appeal the High Court decision directly to the Supreme Court, thereby leap-frogging the Court of Appeal. This was to save time and costs. Taylor approached then barrister David Kirkpatrick to strengthen the EDS legal team. As Kirkpatrick recalled,

> the phone went and it's Gary saying 'would you act for us if we can get to the Supreme Court'. And I said of course I would, absolutely. However, at that time, I was under consideration for appointment as an Environment Judge and as Chair of the Auckland Unitary Plan Panel. Gary was concerned and said 'don't become a judge before the hearing'. As it turned out, it was a close-run thing, and this was the last case I ran as a barrister.

> I had several meetings with Gary, Rob and Nicola and at that point we began to analyse the High Court decision. We had a lot of discussion on how we should frame the case if we got leave. I was conscious that we couldn't just rerun what happened in front of the board of inquiry and the High Court. We had to have something to catch the interest of the appeal judges. So I said to the EDS team 'in front of the Supreme Court on second appeal you have got to have a really sharp dagger not a broad sword'.
>
> I had not gone to the Salmon lecture the year before where Chief Justice Dame Sian Elias presented on 'Righting Environmental Justice', but I obtained a copy of her paper and read it very carefully. I thought wow she is giving a really clear message to those practising in this field. So I said to Gary, Rob and Nicola 'I don't know if her views are shared by others on the Supreme Court but she will be presiding and she is saying get right back to the law. So if we are going to run a legal argument we need to get back to the text, get right back to the principles that underpin the RMA. I thought we needed a really focused, limited, black letter law argument.'

In a nutshell, EDS's legal argument was that policies in the New Zealand Coastal Policy Statement required decision-makers to 'avoid adverse effects of activities' on areas of outstanding natural character and outstanding natural landscapes in the coastal environment. And the RMA required regional plans to 'give effect to' the Coastal Policy Statement. EDS thought that 'avoid' meant avoid in the normal sense of the word, that is, not letting such adverse effects occur.

EDS's application to appeal directly to the Supreme Court was strongly opposed by Crown Law and New Zealand King Salmon but was ultimately granted by the Supreme Court. Save Our Sounds also got leave to appeal and the two cases were run in tandem. Five judges presided over the cases.

As Taylor recalled after the hearing,

> the Supreme Court case reinforced for me the old adage that less is more when it comes to submissions. David was deliberately brief and focused in hard on the key arguments. I believe others (with apologies to them) went on at unnecessary length. The members of the court were really onto it, as you might expect. I felt pretty good about our position at the end and somewhat vindicated at the decision to pursue the case. It was clear that we were onto something important.

Kirkpatrick recalls,

> the greatest worry I had during the hearing was that the Save Our Sounds case, which ran first, would never finish. We had four days set down for the hearing and it was getting into the third day and that case was still going. Finally, with a hurry up from the court, we got the chance to say what we wanted to say. If anything, the shortage of time reinforced my view that we really had to get quite focused on the core point of 'give effect to'. However, what we found when we ran that argument was that 'give effect to' was not that exciting so 'avoid' became the big thing that went back and forth between the judges.
>
> As I debated the legal points with the judges, I sensed that the Chief Justice and

> Justice Glazebrook were with us. Then Justice Young said 'if you say you have to avoid then you have to avoid all adverse effects. You couldn't have anything.' I said 'may it please your honour but you avoid what is inappropriate'. He said 'inappropriate, what does that mean, a navigation buoy? What about a lighthouse, it's a human-made structure?'. 'Yes,' I said, 'but it saves people from being killed.'
>
> He was concerned that the approach might be overly binary and that avoid simply means avoid and you can't have your cake and eat it too. If you can't have a marine farm you can't have anything else. You can't have baches and you can't have yachts and all the various things that are all through the Marlborough Sounds. I said some things are appropriate and some things are inappropriate, and in the end the majority agreed.

When the Supreme Court decision was released in April 2014, it was a landmark moment for EDS and the resource management profession more generally. EDS had won the case which turned more than 20 years of resource management practice on its head. It was the first time the highest court in the land had considered New Zealand's prime environmental legislation, the RMA. As the Supreme Court commented in its decision, the issues 'had not previously been considered by this Court and [the decision] has the potential to affect all decisions under the RMA'. And so it did.

For David Kirkpatrick, who by the time the decision was released was an Environment Court judge, the decision was a huge relief. For lawyer Nicola de Wit, the King Salmon case was the highlight of her time at EDS. She recalls, 'It was just an awesome experience. What amazed me the most was how quickly the Supreme Court justices grasped everything and asked very good questions. I still get comments like, "oh you're the one from King Salmon".'

In a majority decision, the Supreme Court took a fundamentally different approach to the RMA than had been applied previously. As far as the court was concerned, the use of the word 'avoid' in the Coastal Policy Statement meant what it said, it meant 'to not allow' or 'prevent the occurrence of'. It provided something in the nature of a bottom line.

The court roundly rejected the overall judgement approach, which it found to be inconsistent with the elaborate process required to issue a national policy statement. 'If there is no bottom line and development is possible in any coastal area no matter how outstanding, there is no certainty of outcome.' The court went on to explain:

> The 'overall judgement' approach allows the possibility that developments having adverse effects on outstanding coastal landscapes will be permitted on a piecemeal basis, without a full assessment of the overall effect of the various developments on the outstanding areas within the region as a whole. At its most extreme, such an approach could result in there being few outstanding areas of the coastal environmental left, at least in some regions.

It was a detailed and wide-ranging decision that went much further than the matter at hand, a proposed salmon farm in Port Gore. It drilled right into the heart of how the resource management system under the RMA was intended to operate. It sent a shock

Even though EDS beat New Zealand King Salmon Limited in the courts, it still maintained a good relationship with the company. Shown here is an EDS field trip in 2019 to one of the company's new salmon farms.
RAEWYN PEART

wave throughout the resource management community. The proposition that 'avoid' meant avoid, and that firm environmental bottom lines could be set in policies and plans and could not be undermined by development proposals, was nothing short of revolutionary. As EDS policy director Raewyn Peart observed at the time, 'if you dreamt of one key decision during your entire resource management legal career this was it. I was so chuffed that EDS finally managed to get the RMA before the Supreme Court.'

Subsequently, as the dust settled, there were many claims that the decision was completely unworkable, that development would be stopped in its tracks. Pressure was put on government to weaken the Coastal Policy Statement and remove the word 'avoid'. There were attempts in subsequent cases to confine the decision to it factual situation (i.e., to salmon farms in the Marlborough Sounds) and to read down the implications of its findings. EDS had to remain vigilant to stop the decision being watered down. It also worked hard to ensure that the case was reflected in council planning documents. That work is ongoing.

—ooo—

IN THE LATE 2000s, aware that most regional policy statements in the country were due to be reviewed, EDS started work on producing a document to assist regional councils in preparing them, as well as parties who were submitting on the documents. Senior planner Peter Reaburn helped EDS policy analyst Raewyn Peart put the guide together and the *Guide to Strengthening Second Generation Regional Policy Statements* was published in 2011. The guide emphasised the importance of second-generation regional policy statements being much clearer and more directive than the first-generation plans, which had typically been high-level, vague documents that said little more about managing development than the RMA itself.

In order to help cement a new approach within the system, EDS decided to submit on all regional policy statements that came up for review. Under a 2005 amendment to the RMA, all regional and district plans were required to give effect to regional policy statements. EDS therefore hoped that if it was successful in strengthening these higher-level documents, there would be less need for the society to get involved in the hundreds of plans that were developed within their ambit. In addition, many councils were not up to date with the implications of the strengthened New Zealand Coastal Policy Statement and the King Salmon Supreme Court decision.

It was long, hard and unglamorous work. EDS lawyers and expert advisors attended numerous mediations and hearings around the country to argue over the detailed wording contained in the documents. Often the regional councils dominated by the rural sector were less than sympathetic to EDS's submissions. As planner Dave Serjeant recalls,

> I got involved in EDS's Southland regional policy statement submissions. We were pitching at the usual customers on the other side, Federated Farmers and the council, which was farmer-oriented. It was a bit of a steep road to travel there. The King Salmon case was going through at the time and we were bringing the Southlanders up to speed on where the case had got to.
>
> [EDS lawyer] Madeleine [Wright] had a hard job. She had to go around the regions telling people 'you need to get up to speed with what the King Salmon decision means. Irrespective of what your councillors might think, this is what the law is now. Your regional policy statement, being the predominant planning document in the region, needs to be consistent with that, especially in its interface with the Coastal Policy Statement.' It always comes down to the technical wording and that was where Maddie was great. She said 'these are the words you need' and she was spot on all the time.
>
> I remember being at an Environment Court mediation with Ross Dunlop and we were having a break. We were going out into the corridor and a couple of the other parties were there. Ross said to them 'you need to listen to what Maddie and Dave are saying, as what they're saying is right. You may think it's not what your employers want to hear but listen to what EDS is saying.' Ross is a straight-shooter and he could see the writing on the wall.

Most of EDS submissions on regional policy statements were settled by mediation, and EDS was successful in inserting more directive language and clear bottom lines for important environmental matters. As Madeleine Wright explained, 'EDS had a good approach of focusing on a narrow range of matters supported by evidence. We made a lot of ground getting consistency across the country. For example, for offsetting and compensation we brought increased knowledge and understanding of the terms, and we clearly defined "offsetting".'

One regional policy statement that didn't settle through mediation was the one for Otago. That generated controversy about provision for port expansion. It also generated debate about provisions for offsetting in relation to mining. These ended up in the High Court where EDS and Forest and Bird presented a joint case represented by barristers Sally Gepp and Madeleine Wright. Shona Myers gave supporting biodiversity evidence.

The environmental groups argued that biodiversity offsetting and compensation should only be provided for if there were clear parameters or limits that would ensure that biological diversity was maintained. Mining company OceanaGold, the Crown and the Queenstown Lakes District Council argued that there should be no limit on the circumstances in which compensation could be applied.

In the end, EDS was successful in getting limits inserted into the regional policy statement. As EDS in-house lawyer Cordelia Woodhouse commented, 'it was a significant case, setting limits for offsetting and compensation. The decision established case law to the effect that, in some instances, indigenous biodiversity is so rare or irreplaceable that offsetting and compensation is not appropriate. That is a very important part of offsetting rules in plans.'

IT WAS NOT LONG before there were threats to the King Salmon jurisprudence. One was in relation to the landscape policy provisions applying to Man O'War Station on Waiheke Island in Tīkapa Moana Hauraki Gulf. Plan Change 8 to the Auckland regional policy statement had applied a new set of maps showing outstanding natural landscapes in the Auckland region. The case related to the landscape mapping on Man O'War Station on the eastern end of Waiheke Island as well as on areas of Ponui Island, which had the same owner and operated as a farm. Rob Enright and Madeleine Wright represented EDS, which appeared as an interested party in support of Auckland Council.

In appeals to the High Court and Court of Appeal Man O'War Station argued that as the King Salmon Supreme Court decision had given stronger protection to outstanding natural landscapes on the coast, the threshold for a landscape to be classified as 'outstanding' should be set 'at the very highest level'. This was on the basis that the outstanding natural landscapes proposed by the Auckland Council would inhibit ongoing use and development of the land.

Man O'War Station also argued that the value of a landscape should be assessed within the national context, which would then set a higher bar for regional landscapes to be designated 'outstanding'. This was because the landscapes on Waiheke Island would then potentially be compared to those in places like Milford Sound or the Wakatipu Basin and, in that context, would be less likely to rank highly.

Fortunately, both the High Court and the Court of Appeal dismissed Man O'War Station's appeals, thereby upholding the letter and spirit of the Supreme Court King Salmon decision. They found that an assessment of whether a landscape was outstanding was 'essentially factual', being determined by a technical assessment typically undertaken by landscape architects, and should not be affected by the restrictions that might apply once the land has been identified as such.

A MORE RECENT challenge to broader application of the King Salmon case has been fought out in the context of port development in the Otago Harbour. The matter arose out of the Otago regional policy statement and its provisions applying to Port Chalmers and Port Dunedin. The matter at issue was whether ports

One of the cases that followed King Salmon focused on Man O'War Station on Waiheke Island. The owners unsuccessfully argued that because of the stronger protection given to outstanding natural landscapes after the Supreme Court decision the criteria for identifying them should be tightened.
RAEWYN PEART

should be excepted from the requirement in the New Zealand Coastal Policy Statement to avoid adverse effects on specified outstanding areas.

EDS appealed the Otago Regional Council's decisions on the proposed regional policy statement to the Environment Court on the basis that they failed to require Port Otago to avoid such effects. Port Otago and the Otago Regional Council responded by arguing that the port should only be required to avoid such effects 'as far as practicable'. Otherwise, they contended, the port might be required to shut down. The Environment Court sided with this view, so EDS appealed to the High Court. EDS thought that the matter of principle at stake was too important to be left hanging. In the society's view, the Environment Court's approach had fundamentally watered down the impact of the Supreme Court King Salmon decision and it was the thin end of the wedge.

The case was heard in the High Court in June 2019. This time Douglas Allan represented EDS; Forest and Bird was there in support. Marlborough District Council also became a party to the proceedings, given the broader ramifications of the matter for other councils. The High Court characterised the underlying issue behind the case as being 'what is essentially a trade-off between broad commercial interests involving development and the wellbeing of the community on the one hand and preservation of the coastal environment on the other'. It carefully reviewed the Supreme Court's King Salmon decision and found the Environment Court had not applied the correct legal test. EDS's appeal was upheld.

But the matter did not rest there. Port Otago then sought leave to appeal directly to the Supreme Court and succeeded. It was challenging the application of the King Salmon decision in the very court that had decided it. After a hearing, and an 18-month-long wait, the Supreme Court decision was issued in August 2023. This effectively kicked the ball down the road and left it to the regional policy statement to resolve conflicts between the

Another challenge to the King Salmon Supreme Court decision was made in relation to Port Otago. The port company argued that applying the strict rule of avoiding adverse effects on outstanding natural landscapes could require the port to shut down. The Supreme Court declined to overturn its earlier King Salmon decision, leaving it to the regional policy statement to resolve conflicts between the needs of the port and landscape protection.
RAEWYN PEART

landscape and port policies in the Coastal Policy Statement. But of enormous relief to EDS, the decision did not overturn the King Salmon decision.

—ooo—

THE RESOURCE Management Act (RMA) has been the bread and butter of EDS's work since the society was re-stablished in 1999. However, over the years it became apparent that, despite its promise, the legislation was not serving to protect the environment as intended. Part of this was due to the application of the overall judgement approach that the King Salmon decision had overturned. But there were also other deficiencies in the resource management system that EDS sought to rectify.

The idea of establishing an environmental protection authority was first floated in the National Party's 2006 *Bluegreen Vision* discussion paper and it was included in National's 2008 election policy. New Zealand was one of the few countries in the world that did not have a national environmental regulator.

Sensing the growing political appetite for establishing such an agency, EDS initiated a piece of work to investigate models of environmental protection agencies overseas and develop recommendations for such an agency in New Zealand. Gary Taylor and Raewyn Peart visited environmental protection agencies in Sweden, Denmark, Scotland and the Australian states of Western Australia and Victoria. The work was supported by the New Zealand Law Foundation.

The resulting report, *Improving Environmental Governance: The Role of an Environmental Protection Authority in New Zealand*, was released in June 2009. One of the important recommendations, which was subsequently taken up by the government, was that the new agency should be at arms' length from government and have its own independent board. This was the structure that was adopted when the new agency was established as a Crown entity in 2011. EDS is now further considering the future role of the Environmental Protection Authority in a reformed resource management system.

DURING THE 2010s, the clamour over the deficiencies of the RMA heightened and 'RMA-bashing' became a popular sport. This was particularly in the context of the development of urban housing and associated infrastructure for a fast-growing population. The development community wasn't happy with the legislation and it was clear from what EDS and other environmental groups had been saying in the media that they thought the RMA was not working for the environment. As a result, the New Zealand Council for Infrastructure Development (now Infrastructure New Zealand), Employers and Manufacturers Association (Northern) and the Property Council of New Zealand jointly funded EDS to undertake an evaluation of the environmental outcomes of the RMA. EDS policy analyst and former planner Dr Marie Doole (née Brown) led the work.

It was a difficult assignment. Although the Minister for the Environment had a statutory function to monitor 'the effect and implementation' of the RMA, little monitoring or other quantitative research had been undertaken. The RMA had been severely tampered with over the years by successive governments, with some 21 amendments, and it was now much longer and less coherent than its original form.

As part of the research, Doole undertook interviews with 48 key people who had in-depth experience with the resource management system. When asked if they thought there had been environmental decline or improvement since 1991, when the Act came into force, 81 per cent thought there had been a decline. Only one interviewee considered that the goals of the RMA had been achieved. It was a sad indictment on the performance of the once novel and ambitious system.

EDS's report *Evaluating the Environmental Outcomes of the RMA* was released in June 2016, and it confirmed what many had suspected, that the RMA had not met expectations in terms of achieving environmental outcomes. The report put that largely down to poor implementation, but a wide range of other factors also contributed. It stated, 'There is a significant gap between the statutory aspirations of the RMA and the outcomes actually achieved. All of this information in combination paints a picture of a world-leading piece of legislation implemented relatively poorly from day one.'

The Act had been particularly weak in managing cumulative effects and addressing longer term strategic issues. The EDS report was influential as this was the first time that the environmental performance of the Act had been reviewed in a rigorous manner.

The report was followed soon after by an investigation into compliance monitoring and enforcement which the society suspected was one of the weakest aspects of the entire resource management system. The project was again led by Marie Doole. Prior to undertaking a PhD and joining EDS, Doole had been a compliance and monitoring officer at North Shore City Council and so had seen first-hand the challenges in this area. To help inform the project, Doole sent a questionnaire to all 78 councils around the country and interviewed more than 200 people involved in environmental compliance activities.

As Doole recalls,

> most people didn't think that CME [compliance monitoring and enforcement] was an issue. But I knew the sector was waiting for a report that pulled everything together. The process of developing it was tortuous because I wanted it to be heavily empirical and not based on assertions and assumptions. I discussed the approach with several people and said I planned to survey councils on their enforcement practices. They said 'councils will never reply.' But they did, all 78 of them, and I had the most comprehensive survey of local government that had ever been done in the compliance space. The Department of Conservation also engaged well with me and even started to fix things up while I was doing the research.

What Doole found was profoundly shocking for many. There was very patchy and frequently poor compliance performance as well as political involvement by councils in decisions on whether to prosecute. Compliance monitoring and enforcement had been under the radar until the report, *Last Line of Defence*, was released in February 2017. It was clear that all other efforts around planning and consenting would come to nought if compliance with the rules was not enforced. Doole was invited to join a roadshow about her findings hosted by the Resource Management Law Association. Her work prompted significant change.

As Doole reflected, 'more than any of my publications, *Last Line of Defence* keeps coming up as there is no other report like it. I still remember the launch, which was during my last week at EDS, and Gary got all emotional. I then went off on maternity leave and watched all the media. My daughter was born a few weeks after the report was released. It was great to leave EDS with such a good finish.'

—ooo—

FOLLOWING ON FROM the review of the environmental performance of the RMA, EDS formed a coalition with business entities to push for its reform. The advocacy was undertaken under the banner of 'Resource Reform NZ'. The alliance between business and environmental sectors gained political attention and helped create the political opportunity for reform.

As Environment Minister David Parker observed,

> through bringing together the Property Council, Infrastructure NZ and the EMA [Employers and Manufacturers Association], EDS was able to broker an understanding that everyone's needs could be met within a framework that had better outcomes, both developmentally and environmentally. That led to an unblocking of future resource management reform in New Zealand. If EDS hadn't done that, I don't think we would be where we are now. It led to the 'Randerson Panel' [the Resource Management Review Panel which was established by Minister Parker to provide recommendations on resource management reform] which is leading to a better outcome.

Alongside this advocacy, EDS was keen to strongly influence what any future reform might look like. To do this it mounted the biggest policy project it had undertaken to date; a fundamental review of the entire resource management system. Dr Greg Severinsen headed the work and brought to it his keen intellect and ability to undertake systems-wide thinking. Raewyn Peart worked closely with Severinsen for much of the project.

The work started in 2017 and continued on into 2020. It was funded by the New Zealand Law Foundation, the Michael and Suzanne Borrin Foundation and others, and had three phases. It took a first-principles look at how the whole resource management system operated, and provided perspectives on how it could be improved for the coming decades. The first phase was about defining the system, breaking it down into themes and providing various reform options. This produced an extensive synthesis report which put forward three overall models for what a future system could look like. The report won the Resource Management Law Association's best publication award.

As Severinsen recounted,

> It was really fascinating to see, over the course of the year and a half doing the EDS project, how much things changed. It went from being exploratory, and not going anywhere fast, to there being a real sea change in the way government and stakeholders were thinking about deep reform. A crucial turning point, when government started listening, was when we developed partnerships along the way with business. We were lucky to have a reform-minded government that wanted to do something and leave a legacy.

Stephen Selwood (left), CEO of the NZ Council for Infrastructure Development, and Kim Campbell, CEO of the Employers and Manufacturers Association (Northern), were part of the EDS-business resource reform coalition. They are shown here participating in an EDS panel on RMA reform. RAEWYN PEART

A series of working papers were produced during phase one of the project to float ideas and prompt discussion. Working paper three profiled innovative thinking from a range of external experts including Dr Robert Joseph on the Treaty of Waitangi, tikanga Māori and the RMA; Professor Tim Hazledine and Dr Tim Denne on economics and resource management; Dr Theo Stephens on economic instruments; and Dr Marie Doole on compliance and evaluation.

The project's second phase produced another extensive report that put forward a single preferred model for change and offered thoughts on a pathway to get there, recognising that not everything needed to happen at once.

The third phase looked at the urban context, emphasising the need for fundamental reform of how New Zealand's cities are planned and managed. As Severinsen observed, 'the project kind of ballooned into much bigger things

After completing degrees in science and environmental law from Auckland University, **Dr Marie Doole** (née Brown) worked as a compliance and monitoring officer and then environmental policy advisor for North Shore City Council for close to five years. She then commenced work on her PhD at the University of Waikato, which she completed in late 2013 prior to joining EDS as a senior policy analyst. Her doctoral work focused on ecological compensation.

COURTESY OF MARIE DOOLE

Doole worked for EDS from August 2013 to March 2017. During that time, she authored and co-authored many notable reports, including *Vanishing Nature*, which chronicled the loss of biodiversity and reasons for its demise; *Pathways to Prosperity*, which explored how biodiversity could be better safeguarded in development processes; *Evaluating the Environmental Outcomes of the RMA*, which assessed the environmental performance of the RMA, and *Last Line of Defence*, which critiqued environmental compliance monitoring and enforcement.

As Doole recalls, 'my role at EDS gave me a lot of autonomy and the ability to define problems on my own, rather than having a project given to me. The strength of EDS was its ability to be responsive to trends, concerns and emerging issues. EDS often set the framing for the conversation and brought things to light that others might rather have hidden. It sees the world in a multi-faceted way and has access to the right people. The influence of EDS is tentacular.

'The great thing about EDS is that it is a really good team. EDS is a magnet for the good, kind-hearted people of the world. It has a reliance on graduates and a revolving group of staff come through. When I was there, I worked alongside some pretty amazing people, the likes of Rob Enright, Devon McLean and Rob Fenwick. The society is incredibly well connected. It is also a nursery of capability and has a formative effect on people. EDS always looks for a solution. The attitude is, we have to fix it somehow, there has got to be an answer. It gives you that sense of optimism.'

After leaving EDS, Doole worked for environmental consultants, The Catalyst Group, for just over three years, and then became a principal advisor on compliance for the Waste Operations Team in the Ministry for the Environment.

as the government took up the reforms. In phase two of the work we got to pin our flag to the mast and say "this is what we want". We got to put EDS in a really strong position to influence what was happening in a once-in-a-generation-type process.'

The project involved diverse topics including international law, legal principles and environmental ethics, legislative design, governance and institutional structures, participatory arrangements, and legal and economic tools. It looked at what was not working well with the current system, but also at what innovative and different approaches could be taken in a future that would look very different from the present.

The political momentum for change continued to build as the project progressed.

The last phase of EDS's resource management reform project focused on the urban context (such as the Auckland CBD shown here) and emphasised the need for fundamental reform of how New Zealand's cities are planned and managed. RAEWYN PEART

During phase two of the work the government established an independent panel, chaired by the Hon Tony Randerson QC, to look into fundamental reform of the resource management system. Raewyn Peart was appointed to the panel and the EDS project was specifically identified by the Minister for the Environment as an input to be considered. In mid-2020, the panel produced a well-considered and thoughtful final report that in many ways echoed the key conclusions of the EDS project.

Deep reforms of the resource management system are currently underway, and EDS's input laid some of the groundwork for the generational change expected. Over the course of 2021, EDS provided detailed advice to the Ministry for the Environment on environmental limits, national direction, strategic spatial planning, landscape and connections with the marine environment, to inform policy development and drafting, with Greg Severinsen leading much of the analysis.

EDS continued to engage deeply in the reform process and lodged comprehensive submissions on the two Bills that were introduced to Parliament. The Natural and Built Environment Act and Spatial Planning Act were both given royal assent on 23 August 2023. It was potentially the end of one era and the beginning of another, depending on what future governments do with the new legislation.

—ooo—

EDS HAS ALSO been closely involved in the process to develop a national policy statement for indigenous biodiversity under the RMA. Despite having nearly 4000 species threatened with extinction or at risk, until very recently, New Zealand lacked any national policy to provide direction to councils on how they should manage development impacts on biodiversity.

The need for a national policy statement on indigenous biodiversity to protect native flora and fauna on private land had been recognised since at least 2000. In January 2011, National's Environment Minister Nick Smith and Conservation Minister Kate Wilkinson released a proposed policy statement for public comment. But as the measures to protect biodiversity on private land became

Dr Greg Severinsen graduated from the University of Canterbury with a Bachelor of Laws (Hons) and Bachelor of Arts in history in 2009. He then joined Simpson Grierson to work as a junior lawyer but after a while found himself thinking 'it's a bit frustrating just focusing on the needs of clients rather than thinking more broadly about why we might be doing something'. So after almost three years he decided to leave private practice and embark on a PhD in law at Victoria University. His thesis investigated how the RMA dealt with carbon capture and storage.

When Severinsen finished his doctoral studies, a position opened up at the Ministry for the Environment in early 2016, and he jumped at the chance to work on early-stage thinking about reforming the resource management system. But the political appetite for reform ebbed and flowed and the work didn't seem to be going anywhere. So when Severinsen saw an opening at EDS to do similar work with independent funding it seemed right up his alley.

He joined EDS in June 2017 and is currently a policy director for the society. He led the society's very influential resource management reform project which resulted in new legislation being passed by the government in 2023. Reflecting on his time at EDS, Severinsen observed, 'it's been quite a different experience compared to my previous professional lives as an academic, lawyer and policy analyst for government. It gives you more freedom and creativity, the ability to publish and get ideas out there, and to talk to interesting people without fear or favour. The small, close-knit team I am working with makes it an environment where you can think creatively and express ideas. And given the mana that people like Raewyn [Peart] and Gary [Taylor] have, our ideas really have an impact.'

Severinsen lives and works in New Plymouth and finds it 'nice to be able to work in the kinds of places that inspire you. Often I will work in a café that looks out over the ocean. It's not just being able to work from the regions but also being able to work in places where you feel connected with the nature you are trying to protect. You can get wound up in the technical detail of the work, but if you take a step back and look at what EDS is achieving, 20 years from now it will be part of history.' He is shown here visiting the OECD in Paris during an EDS international research trip.

embroiled in controversy, it went nowhere.

In 2017, the Labour-led government decided to have another go. Aware of the controversy surrounding the topic, Environment Minister David Parker decided to set up a collaborative group to see if the protagonists could agree on a way forward. Members included representatives from Forest and Bird, Federated Farmers, the Forest Owners Association, infrastructure industries, the Iwi Chairs Forum and EDS.

Former EDS lawyer Madeleine Wright represented EDS on the group. She observed,

> It was a really intensive process. A lot of work went into trying to get agreement. There may be some outstanding positions on some things but the fact there is agreement between EDS, Forest and Bird and Federated Farmers on avoidance is huge. It was a very gruelling process. Over a 12-month period we were in Wellington

It took until 2023 for a policy statement on the protection of indigenous flora and fauna on private land to be finalised. Shown here are patches of native vegetation on farmland in Akaroa. RAEWYN PEART

> every two weeks for long meetings. The meetings were intense. There were huge amounts of work between meetings with lots of technical information to digest. It felt like a big success and then the government did nothing. So it sat languishing for some time.

The collaborative group reported to the minister in October 2018 and then little seemed to happen. It was not until June 2022 that a draft of the national policy statement was released for public comment. EDS engaged Wright to draft the society's submissions. The policy was finally promulgated in July 2023.

All the national policy, including the New Zealand Coastal Policy Statement and the National Policy Statement on Indigenous Biodiversity will be drawn together into a new National Planning Framework to be promulgated under the Natural and Built Environment Act. Greg Severinsen has been appointed to a Ministry for the Environment advisory group for the development of the new planning framework.

—ooo—

A MORE RECENT project for EDS has been reforming conservation law. Marie Doole headed the society's first piece of work in this area, which resulted in the publication of *Vanishing Nature* in 2015. This explored the biodiversity crisis facing New Zealand and the reasons behind it. The report identified the three key drivers of biodiversity loss: markets not valuing biodiversity; weak regulation; and weak public pressure for conservation. The report set out a wide range of solutions that could be practically implemented.

This was the first project Doole undertook for EDS. Retired Department of Conservation scientist Dr Theo Stephens, with his inside knowledge of the system, agreed to step in and help out. It was a baptism by fire. As Doole recounts:

> we put a draft out and got fairly brutal feedback. It was quite crushing. Fortunately, we had the flexibility to circle back and regroup and got a much better product from it. Six years on *Vanishing Nature* is still a university text, so it is going okay. It needs review now, but I would say 50 per cent of the recommendations have been implemented so far, so that's good.

But even given such improvements, it became clear that much of the legislation was outdated, conservation planning processes had become bogged down, and there was a need for a rethink. This was particularly highlighted with

The first project Marie Doole undertook for EDS explored the biodiversity crisis facing New Zealand and the reasons for it. The results were published in a report titled *Vanishing Nature*. Shown here is Doole at the launch of the report, with co-author Dr Theo Stephens to the right and her former PhD supervisor Professor Bruce Clarkson in the centre. RAEWYN PEART

the exponential growth of the tourism industry prior to Covid-19 and much greater pressures being placed on the use of conservation land.

In response, EDS decided to take another look at the system, supported by the Department of Conservation. Work on the project started in May 2020. Dr Deidre Koolen-Bourke was the prime researcher, supported by Raewyn Peart. Koolen-Bourke had recently completed her PhD and brought her considerable research skills to the task. The pair decided to investigate several case study areas, so went out on the road to interview people on the West Coast, Otago and Nelson. They also undertook several workshops with DOC staff. All in all, more than 50 interviews were completed.

Koolen-Bourke found the field work fascinating. 'Most of my previous work had been desktop rather than getting out to talk to

In order to investigate the conservation system, EDS policy director Raewyn Peart and senior policy researcher Deidre Koolen-Bourke headed out into the field to interview people. One of the places they visited was the Ōkārito lagoon, to investigate the management of tourism concessions. Shown here is Koolen-Bourke at the lagoon. RAEWYN PEART

people and finding out what the issues really are. What is really different about EDS is the amount of talking with stakeholders and government departments and the interviewing process. It gives you a much deeper understanding of how the issues play out.'

The resulting report, *Conserving Nature*, was launched at the 2021 EDS conference in Christchurch by Conservation Minister Kiritapu Allan. The publication was designed to look similar to *Vanishing Nature* and it proved highly popular, winning the Resource Management Law Association's Publications Award that year. It was the first time that a description of the entire conservation management system, and issues with its operation, had been brought together in one digestible publication. EDS is now undertaking stage two of the work — to look at options for reform. This includes the reform of the Wildlife Act, the conservation management planning system and the conservation system overall.

—ooo—

COURTESY OF DEIDRE KOOLEN-BOURKE

Dr Deidre Koolen-Bourke had wanted to be a conservation biologist when she was at school. At the University of Auckland she subsequently completed a conjoint Bachelor of Arts in anthropology and Bachelor of Science in 1994 and then a Bachelor of Laws (Hons) specialising in public, environmental and animal law in 2004. In 2007 she completed a Master of Laws (Hons) in environmental law and was admitted to the bar the following year. After a break in studies to have a family, Koolen-Bourke returned to study in 2012 to undertake a PhD. She completed this in 2019, shortly before joining EDS in March 2020 as a senior policy researcher.

Since working for EDS, Koolen-Bourke has completed reports on the conservation management system, *Conserving Nature*, and the role of science in the freshwater policy process. She has also contributed to EDS's landscape and resource management reform projects. Koolen-Bourke is the lead researcher for the second stage of EDS's conservation law reform project.

NEW ZEALAND'S SOLE aluminium smelter is located at Tiwai Point in Motupo-hue Bluff and is operated by New Zealand Aluminium Smelters Limited. Around 80 per cent of that company is owned by Pacific Aluminium (New Zealand) Limited, a wholly owned subsidiary of Rio Tinto Limited, which is an Anglo-Australian company producing metals, including aluminium, worldwide. The other 20 per cent is owned by a Japanese company, Sumitomo Chemical Company Limited.

A by-product of the aluminium production process is a material called 'dross'. Dross can be further processed to recover aluminium and one of the outputs of that process is 'ouvea premix'. This in turn can be used in fertiliser manufacturing and steel production. Ouvea premix contains a number of substances, including aluminium oxide and aluminium nitride. Aluminium oxide naturally absorbs water from the air and aluminium nitride reacts with water to produce toxic ammonia gas. This means that ouvea premix needs to be stored away from air and water.

New Zealand Aluminium Smelters had adopted a practice of contracting out the processing and recycling of ouvea premix. One of the companies contracted to do this, Taha Asia Pacific, went insolvent in 2016. On doing so it left 10,000 tonnes of ouvea premix stored in an old papermill located next to the Mataura River in the small Southland town of Mataura. The river was known to be subject to flooding; in 1978 the paper mill buildings had been flooded to a depth of up to 2 metres. The river flooded again in 1980, 1984, 1987 and 1999.

Taha Asia Pacific moved the ouvea premix waste into the paper mill buildings during 2014. It had initially planned to establish a mineral processing plant at Mataura to produce mineral fertiliser, but this idea was later abandoned in favour of a new site in Waihōpai Invercargill.

Around 10,000 tonnes of ouvea premix from the Bluff aluminium smelter was stored in the disused Mataura paper mill, shown here, which was subject to flooding. If it becomes wet, the premix can release a toxic ammonia gas. EDS became involved in the successful court proceedings that resulted in New Zealand Aluminium Smelters eventually removing the premix and safely storing it.
MATAURA MUSEUM

The company applied for retrospective consent for the ouvea premix to be in Mataura and this had been granted by the Gore District Council for two years, giving time for the waste to be removed to the new Invercargill processing plant. However, before this was done, the company became insolvent.

New Zealand Aluminium Smelters subsequently reached an agreement with the government and the Gore District and Southland Regional councils in March 2018 to provide $1.75 million towards the $4 million cost of removing the material, which was to be undertaken over six years. But by the time of the flood in 2020, only around 500 tonnes of the estimated 10,000 tonnes stored at Mataura had been removed. The flood brought matters to a head.

In February 2020 there was a downpour, and some of the floodwaters reached the Mataura paper mill, which was next to residential properties and close to a childcare centre and school. The prospect of toxic ammonia gas filling the town if the ouvea premix got wet prompted widespread evacuations. There would also potentially be serious harm to river life if the premix entered the water. Extra sandbags were placed around the paper mill and the flood level reached 250 to 300 millimetres from the top of a protection wall around the building. As Mataura Community Board member Steve Dixon described in the *Otago Daily Times*, 'The dross sits atop wooden pallets, which were just high enough to keep water from reaching the substance.'

In response to the flood, Gore District Council Chief Executive Steve Parry announced that he had negotiated a 'handshake deal' with New Zealand Aluminium Smelters to fast track the removal of the toxic material. But this deal was then reportedly vetoed by the parent company Rio Tinto. A statement issued by the smelter's general manager, Stewart Hamilton, claimed that ownership of the ouvea premix transferred to the Crown following the bankruptcy of Taha Asia Pacific. Under the terms of the 2018 agreement, between the company, the government and councils, the responsibility for its removal lay with the Gore District Council.

Gary Taylor read about the Mataura flood in the news. He believed it was wrong that most of the costs of removing the dross, which had been produced by New Zealand Aluminium Smelters in the first place, was being borne by public agencies and that Mataura residents were now subject to significant risk. The society decided to take up the baton on behalf of the community.

In July 2020, EDS announced that it would be filing declaration proceedings in the Environment Court to determine who was legally obligated to remove the potentially toxic material. In EDS's view, New Zealand Aluminium Smelters should be taking immediate steps to remove the dross promptly to a safe site. Rob Enright acted for EDS on the case, with EDS lawyers Shay Schlaepfer and Cordelia Woodhouse undertaking extensive research.

It was a long and tough process. Woodhouse attended the negotiations with Taylor. She recalls,

> we ended up having 12 judicial settlement conferences by zoom. At some points they were weekly. We would get very close to agreement and then something would happen and the agreement would fall away. Being the applicant in the legal proceedings, we had the litigation track we could go down, and we thought we had a pretty strong case. We also kept in mind that the risk time for Mataura was in summer when the water comes down the river.
>
> The mediation kicked off in September and there were some close calls with the river in November and December. We went into the Christmas period and then we filed amended declaration proceedings. We thought we needed to press on for a hearing to keep pressure on the judicial settlement process so that we didn't keep kicking the can down the road. The community were frightened and there was a mental toll on them as they were worried about gases being released.

In February 2021 an agreement was finally reached to move all the material back to Tiwai Point within a six-week period. The costs of removal were to be shared between New Zealand Aluminium Smelters and the government. As Gary Taylor observed at the time, 'this was a complex negotiation that was time-consuming and challenging. While it's our position that a small environmental group like EDS should not have been required to take on a large company to face up to its environmental responsibilities, we are pleased with the outcome.'

During the mediation process there was a change in personnel at the senior board and management levels of Rio Tinto, the holding company of the majority shareholder of New Zealand Aluminium Smelters. EDS wrote to all the new personnel, as well as key company shareholders, suggesting that the stance of the company in New Zealand was not a good look when it came to corporate responsibility.

This appeared to resonate with them and in May 2021 the company announced that it would cover all the costs of removal of the toxic waste to a safe site at Tiwai Point. The last bags of ouvea premix were removed from Mataura in July 2021. They are now safely stored in water-tight shipping containers. In the end Rio Tinto picked up the entire $6 million clean-up bill.

As Richard Brabant later observed, 'I was hugely proud to see EDS get involved in the Tiwai Smelter issue which government did nothing constructive about — but surely could have. It reminded me of the grassroots way that we operated [during the 1970s and '80s] when people in a town like that would contact us when they had made no progress with the council and had a problem.'

16. Inside EDS

IN APRIL 2021 EDS turned 50 years old. That the society is still operating over half a century after David Williams first established it back in 1971 is testament to the hard work and tenacity of many people. It also reflects the successful combination of litigation, law reform and public education that EDS has undertaken over those years.

OVER THE PAST TWO DECADES, the bedrock on which EDS has survived and thrived has been the energy, commitment and considerable political talents of Gary Taylor. Taylor has effectively drawn in others to support the cause. For close to 15 years (between 2006 and 2020) EDS was run by a tight team of three; Taylor acting as CEO and overseeing fundraising and litigation; Raewyn Peart heading policy work and acting as programme manager for the EDS environmental policy conference; and Fiona Driver organising events and acting as general manager to ensure the society ran smoothly.

The trio had complementary skills. Former EDS lawyer Natasha Garvan observed,

Over the past two decades, the bedrock on which EDS has survived and thrived has been the energy, commitment and considerable political talents of Gary Taylor. Taylor is shown here working at the EDS office in 2013. RAEWYN PEART

> Gary is very to the point. I've seen him be very unapologetic and direct but I appreciate that's who he is. Fiona has been described as the 'mother of EDS'. She was always caring, and looking after people and thinking about other aspects than just the work, which all successful organisations need. She also did all the planning and organisation that went into the events. It was not easy to pull all that together.

While Driver provided the social 'glue', Taylor steered the organisation through the constantly shifting political landscape, and Peart brought an intellectual rigour to EDS's work. As Marie Doole observed, 'the great thing about EDS is that it is a really good team. Gary and Raewyn make up the core of the organisation and other people revolve around that. They are the constant that is really important for EDS.' Taylor and Peart still provide ongoing leadership for EDS, with Driver leaving the society temporarily at the end of 2020, but returning in early 2023.

Over the years, the core team has been supplemented by the many talented lawyers and other professionals who have worked for EDS. Many were recruited as young graduates because EDS could not afford to pay experienced lawyers. After several years of working for EDS, staff were often snapped up by law firms offering higher pay rates. It has meant that the society has trained a whole cadre of young lawyers and many have gone on to successful careers in resource management law.

New staff were rapidly drawn into the tight-knit EDS team. As Elizabeth Edmonds explained, 'when I first got involved with EDS in 2005 it was a team of four. Each person was like this powerhouse of brain power with everybody doing what they were good at. You didn't have somebody who was really good at X slotted in and forced to do Y. Everyone respected one another. The gears all worked incredibly well together.'

The team culture was based on a strong

Previous pages: EDS field trip to Ōtata, Tīkapa Moana Hauraki Gulf, 2023. RAEWYN PEART

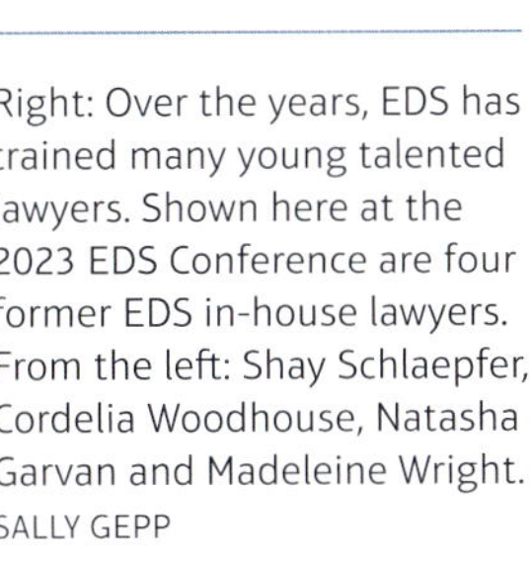

Right: Over the years, EDS has trained many young talented lawyers. Shown here at the 2023 EDS Conference are four former EDS in-house lawyers. From the left: Shay Schlaepfer, Cordelia Woodhouse, Natasha Garvan and Madeleine Wright. SALLY GEPP

Left: The EDS board, staff and supporters on a 2013 retreat at Waiheke Island. EDS held regular board retreats to strategise about where EDS could best focus its efforts. RAEWYN PEART

'can do' attitude, with EDS tackling issues that others found daunting. Kate Storer recalls from her time working for EDS, 'there was a clear ambition to get things done. Almost no challenge was too big. EDS would take on things others would easily dump into the too-hard basket.'

The EDS board has provided important governance oversight of the society's activities with regular director retreats and field trips providing an opportunity to kick around ideas and reflect on where EDS could best focus its efforts. As EDS director Barry Barton observed, 'we kept asking ourselves what our capabilities were and where we could best deploy them'.

The board's finance committee has kept a keen eye on EDS's finances. The committee was for some years chaired by chartered accountant Richard Waddell, followed by engineer Garry Law. The current chair is law professor Barry Barton. Finances were often tight in the early years and caused more than a few sleepless nights. As Barton observed, 'we went through a period in the early days where one poor conference or one costs award could have wrecked us and that was worrying'.

University of Waikato law professor Barry Barton has been an EDS director since 1999. He currently chairs the society's finance committee. RAEWYN PEART

Peart managed the society's accounts during the early years and she remembers that 'we somehow managed to survive month to month. At times the bills exceeded funds in the bank and Gary Taylor would say, "don't worry, I'll defer my fee until funds come in". He always made sure I was paid first. And the funds always did come in eventually.' Today, EDS's finances are on a more even keel. The society has an office in Ponsonby, Tāmaki Makaurau Auckland and a highly motivated team of seven.

EDS's publications have taken on a more professional look over the years. The society moved from spiral to perfect bound, and from black and white to full colour, as printing costs reduced over time. Former photolithographer and now designer Neale Wills, from Wilsy Design, undertook the design of both EDS's community guide to the RMA and landscape policy report in 2004, and he has designed most of the society's community guide and policy publications (over 40), publicity brochures and conference prospectuses since that time. He is an integral part of the wider EDS team.

EDS has worked constructively with governments of very different political stripes. Achieving this has not always been easy. One major pivot point was when the Helen Clark Labour-led Government lost the election in 2008 after nine years in power. As Bryce Johnson explained, 'Gary stood for Labour once, and there was always the risk that EDS would get branded with one political persuasion, which would make it hard to work with opposite governments. But Gary successfully managed to manoeuvre his way through all that. He kept EDS in front on all issues regardless of who was on the Treasury benches.'

On the back of a lot of hard work, EDS gradually built up a solid reputation for being a small but effective organisation. As Kate Storer observed, 'it was extraordinary to be such a small organisation churning out a huge amount of work and having such a huge influence. I was often asked by others how many people were at EDS and I would say five. They would be absolutely shocked as the influence EDS has is so much larger than the sum of its parts.'

—ooo—

SINCE ITS INCEPTION in 1971 EDS has had a strong focus on litigating on behalf of the environment. Litigation is a large part of EDS's *raison d'être* and it is reliant on the support the society receives from the legal and other professions. Former Environment Minister Nick Smith considered that

> the most important role EDS has played over the last 20 years is the presence of someone with horsepower and capacity to be able to legally challenge, whether it be councils or development interests. I

> would know of 20 projects where EDS was not involved in any legal action but where EDS's presence resulted in a different approach from the company knowing that EDS had the legal horsepower to take them on if they strayed from the law.

Environment Minister David Parker reiterated that,

> the court stuff is incredibly important. It's entirely appropriate to allocate resources to fight for good outcomes in the courts. The precedents you achieve there are not influenced by politics. We have a truly independent court system and you get better decisions than through planning departments in councils. So the environmental advocacy through the courts part of what EDS does is absolutely essential.

EDS's litigation programme has become more strategic over the years. As Barry Barton explained,

> EDS has gone through a huge pivot from where it used to be in the '70s and '80s. The Coromandel focus is muted and we are now working at a national level. We are getting involved in places such as the Mackenzie high country and going through everyone's regional policy statements up and down the country. I can remember at a board meeting considering an application for a boat shed somewhere on the Coromandel and we all agreed that we weren't doing that stuff anymore.

Environment Minister David Parker addressing the 2018 EDS conference. EDS

The highlight of EDS's litigation programme would undoubtedly be the 2014 King Salmon case, where the Supreme Court overturned more than 20 years of case law and confirmed that the RMA did contemplate environmental limits. But EDS has achieved many other things through litigating. It has stopped incompatible development on sensitive parts of the coastline. It has helped protect the natural values of places such as the Mackenzie Country and Mapara Valley. And it has reduced pollution of the country's rivers. It has strengthened regional policy statements and plans that set the rules for how the environment is to be managed and held councils to account.

As part of this work, EDS has provided much-needed support to local communities. As former EDS director Stephen Brown noted, 'I think the strength of EDS has always been litigation and not being scared to embark on it. This has given comfort to a lot of community groups and allied conservation organisations that there is a party that supports their views and, if necessary, can turn up to help. That's important as who else would you go to?'

Gary Taylor QSO CNZM was born in Devonport and attended Vauxhall School and Westlake Boys High School. He initially studied law at the University of Auckland, before switching to an arts degree, and completed a Masters degree in history. He then taught English, history and social studies for three years. In 1973, with his artist wife, Mary, he bought a block of land in the Waitākere River Valley and has lived there ever since.

In the mid-1970s Taylor became a current affairs reporter for Radio New Zealand and then editor of the *Forest Industries Review*. In 1978 he was recruited to run EDS. He worked there full-time until 1984 and then part-time until the society went into recess in 1988.

Taylor learnt much from his first stint at EDS. As described in a *Metro* article, 'he turned the coordinator's job which had previously been a dogsbody position that involved looking after the membership and servicing the executive and its projects, into an organiser, negotiator and an advocate, more akin to a managing director than an office manager. During his time there he learnt how to interpret statutes, learnt the intricacies of local government and he came into contact with networks of environmentalists and lawyers. And he learnt how to fly down to Wellington to see the right person to get something done.'

The 1980s was a busy decade for Taylor, when he became involved in local government politics. He gained a seat on the Waitemata City Council in a 1981 by-election and by 1983 was campaign organiser for 'Tim's Team' which captured a majority of the seats on the Waitemata City Council with Tim Shadbolt as mayor. At 39 years of age Taylor was deputy mayor. He was also elected to the Auckland Regional Authority and became chair of its powerful Works Committee. In 1989 he was elected to the Auckland Area Health Board and became chairman.

Over this period, Taylor seemed to rise out of nowhere to occupy some of the most influential positions in Auckland's local and regional government. This prompted *Metro* to label him as the 'Svengali of the west' and as 'The man who rose without trace'.

During the 1990s Taylor worked as a consultant and held directorships of numerous public entities. He played a key role in decisions to build the Britomart rail terminal and to electrify the Auckland commuter rail network.

In 1999 he re-established EDS and has

In the early days EDS struggled to get standing in the courts and lost many cases. But now EDS is asked by the courts to join proceedings and the society rarely loses a case. On the other hand, litigation has become much more time-intensive and costly so EDS has to be more selective and strategic in choosing what matters to become involved in. At the same time, EDS has become bolder in using legal tools to hold decision-makers to account, including seeking declarations where the Environment Court determines a legal issue rather than considering the merits of a proposal.

As former Fish and Game chief executive Bryce Johnson observed, 'EDS is a unique entity in New Zealand, as it focuses very much on litigation. In terms of deeper advocacy with serious substance and law behind it, EDS is way out there. EDS occupies a very significant niche that no one else has occupied in a successful way and maintained over a long period of time. That is a hard thing for a small organisation to do.'

been the CEO of the society ever since. In 2008 he initiated the Land and Water Forum and played a significant role over the next decade in the evolution of the freshwater reforms. More recently, EDS led a campaign for resource management reform.

Reflecting on his career, Taylor observed, 'while at secondary school I was a regular visitor to the headmaster's office, advocating for student rights and complaining about injustices. He called me a bush lawyer and in a way that's what I have been all my life.

'Relocating to the Waitākere Ranges was an epiphany. I grew to love its forests and beaches and my first environmental gig was working with the Waitākere Ranges Protection Society to prevent a rubbish tip from being constructed in the headwaters of the Te Henga wetland. Advocating for nature against powerful vested interests has been a lifelong commitment. So not much has changed really from the schoolboy back in the day.'

Taylor is a frequent media commentator on topical environmental issues. As Bryce Johnson observed, 'Gary was always the very first to get the message out on an issue to get exposure

RAEWYN PEART

and primacy when the media was looking for comment on a government action or proposal.'

Taylor was awarded a Queen's Service Order in 2004 and was made a Companion of the New Zealand Order of Merit in 2019 for services to the environment and resource management.

When reflecting back on his time leading EDS, Taylor observed, 'what a privilege it is to be speaking for the environment in Aotearoa New Zealand. These wonderful islands and their unique biodiversity, landscapes and coastline really are worth fighting for.'

EDS HAS SEEN many iterations of environmental and planning law reform since the early days of the Town and Country Planning Act 1953 and rudimentary water pollution and mining laws. The society has operated throughout much of the lifetime of the Town and Country Planning Act 1977, as well as the Resource Management Act 1991, which was replaced in 2023.

During this time, EDS's policy work has been wide-ranging. Initially focused on matters such as reform of air pollution, water, mining and planning laws, the society's more recent policy work has ranged as wide as landscape protection, coastal management, resource management reform, conservation law reform and oceans reform. In doing the work, EDS has sought to meld lessons from international experience with the domestic context and, prior to Covid-19,

EDS would regularly undertake international study tours to draw on the latest overseas thinking and practice.

As Raewyn Peart remarked,

> I always felt that our international trips gave EDS the edge. It enabled us to bring fresh ideas and perspectives to our work and we could draw on actual lived experience in other countries. Meeting experts in-person is much more valuable rather than just reading the literature. The trips also enabled us to identify engaging international speakers for our conferences.
>
> We also spent a lot of time in the field in New Zealand, visiting sites and talking to people. I always found getting out of the office and seeing what was actually happening on the ground enormously valuable. I particularly liked to meet people who were out on the land and the sea — farmers, fishers, marine farmers and so on — as they had a wealth of knowledge about the environment they were living and working in.

Former Fish and Game chief executive Bryce Johnson worked closely with EDS over many years to improve New Zealand's natural environment, particularly freshwater. Johnson is shown here on a panel at the 2015 EDS conference with Dame Anne Salmond. RAEWYN PEART

As former EDS lawyer Kate Storer observed, 'I learnt that it is really important to see, in person, what you are studying. The field trips showed me that when you are writing about successes and failures of organisations, there are real people involved in working within them. It personalised the issues in a way I hadn't appreciated before.'

'My approach to policy work was very organic,' explained Peart.

> I saw it as a creative process. Some people would start out with a table of contents for the final report, but I would start out with immersing myself in whatever written material was available. This would enable me to see emerging patterns and indications of what the underlying drivers of the problems were. Then I would interview as many people as I could find from a wide range of perspectives. This would provide a wealth of practical insights into the issues. It was only at that point that would I start to conceptualise a structure for the report.
>
> This approach was quite challenging for some funders, who wanted to tie down what the final output would contain before the project started. I would simply say 'trust us' and they usually did. We always tried to over-deliver on the brief and most funders were very happy with what we produced in the end.

Over time, EDS developed a solid reputation for producing high-quality and thought-

Former Environment Minister Nick Smith waiting to address delegates at the 2017 EDS conference.
RAEWYN PEART

provoking policy advice. Elizabeth Edmonds: 'EDS would walk into government and they would truly listen to what they said. There was trust and a bond built with politicians.' EDS researcher Deidre Koolen-Bourke explained, 'they see us as being able to build social licence for change, which governments can't do, as they are not trusted to the extent we are.'

Former Environment Minister Nick Smith:

> Politicians don't do much original thinking. That's not a criticism but is just the nature of having such broad responsibilities. At best, they are idea stealers of which I openly confess to being an active participant. In opposition, most of my ideas that made it into National Party policy in 2008 had their origins in a think-tank like EDS.
>
> The other part that is underestimated is how difficult it is for original big thinking to occur within government departments in an area like climate change or oceans. The problem for government agencies is that they are so siloed, whether it's a fisheries ministry focused on fish, DOC with marine mammals, MBIE with minerals, defence with security, or foreign affairs with the international dimension. It's disappointing how government struggles to ask big-picture questions and develop some of the high-level policy that comes out of organisations like EDS, even with its massive resources.

EDS's policy work is usually ahead of the reform curve, and the government has recognised this. EDS is now often engaged to provide independent reports on law reform issues, such as oceans, the Resource Management Act, the Wildlife Act and the Land Act. While EDS has always tried to push the envelope in terms of providing first principles thinking, the society has also tried to ensure that its recommendations can be implemented in New Zealand's unique bicultural landscape. The highlight of EDS's policy work was the resource management reform project, which not only helped create an appetite for reform within government, but also influenced the nature of the reform itself.

Gary Taylor says that working with Raewyn Peart has been extremely rewarding and he takes credit for talking her out of leaving EDS to do a PhD on more than one occasion. 'When the topic came up it was a relatively easy sell to talk about the reform agenda ahead of us and how critically important keeping her brainpower on the job was. Raewyn has always had an impressive motivation and no one in this country can match her prolific publications outputs. And there is more to come.'

Raewyn Peart MNZM grew up in Hamilton and completed a Bachelor of Social Sciences at the University of Waikato majoring in psychology. After a few years in minimum-wage jobs, Peart went back to study at the University of Otago, completing a law degree and commerce degree (majoring in computer science) in 1987. She was admitted to the bar the following year and joined the litigation team at Kensington Swan in Auckland. After some 18 months there, she joined Russell McVeagh's newly formed resource management team.

In 1992 Peart left Russell McVeagh to work as an in-house lawyer for Brierley Properties Limited, as part of the legal team which sought planning consent for the casino and sky tower development, as well as successfully winning the casino licence itself.

In 1994, Peart headed to the UK, accompanying her then husband on his university sabbatical leave, and then on to Durban, where he had obtained an academic position. Her daughter Tanya was born there. Whilst in Durban, Peart worked for the Council of Scientific and Industrial Research as an environmental policy analyst for four years, while also completing a Masters in Commerce at the University of Natal. In 1999, while still overseas Peart became a director of the newly re-established EDS and she remains a director to this day.

DAVID HARTMAN

Returning to New Zealand in 2001, Peart started working for EDS part-time whilst undertaking other consultancy work. Gradually, as EDS obtained more funds, she increased her hours for the society. She brought her young daughter along to many EDS events, including board meetings and field trips, and Tanya was known as the young 'EDS mascot' for many years.

'I got to the stage in my legal career where I decided that using my skills to make money for clients and the firm was not enough. Although I was well paid and the work was intellectually stimulating, in the end I found it fundamentally unsatisfying. I wanted to use my particular set of skills to achieve positive change in the world. And the biggest buzz I have got from all the work that EDS has done, is when we manage to change something for the better.

'There were dark times when we didn't seem to be making any progress on environmental issues. But often you are making more progress than you think. It's a matter of changing hearts and

minds and that takes time. The important thing is not to give up and to be ready to grab the political windows of opportunity when they emerge.

'The issues have changed significantly over the time I have been involved in EDS. When I first started, EDS's focus was on landscape and coastal development issues. At that time New Zealand still had the reputation of being '100% Pure', and I partly believed in it. But now most of our rivers are polluted and we are facing climate and biodiversity crises which go to the very heart of how we live and create a livelihood as a nation.

'Despite environmental issues becoming more mainstream in recent years, it's still a very hard and slow process to get change. Vested interests will always rise up to oppose it and they are still very influential.'

In 2019, Peart was made a Member of the New Zealand Order of Merit for services to environmental and conservation policy. In 2023 she received the Resource Management Law Association Outstanding Person Award.

Politicians from all the major political parties have generally fronted up to EDS conferences. Shown here at the 2017 EDS conference, which was held shortly before the general election, are from the left: The Opportunities Party candidate Geoff Simmons, Green MP Eugenie Sage and then Environment Minister Nick Smith. RAEWYN PEART

Economist Kate Raworth beaming in from the United Kingdom to address the 2018 EDS conference. EDS

SINCE 2003, EDS has convened more than 30 conferences. They have become high-profile national dialogues on environmental challenges facing the country. They have also served to generate change with, most notably, the Land and Water Forum having its inception during the 2008 'Conflict in Paradise' conference. The conferences have included overseas speakers as well as local experts. Panels have brought together parties from a range of sectors to hammer out solutions to intractable problems. Politicians from all the major political parties have generally fronted up.

Former EDS lawyer Kelsey Serjeant attended several conferences while she was working for EDS and remembers that

> the EDS conferences have definitely been a highlight. It is the best environmental conference and so well attended and it's put on through the hard work of just a few people every year. Raewyn [Peart] and Gary [Taylor] say, 'let's get the best people we can to zoom in from around the world. Let's gun for the best speakers on critical issues.' Every year it is highly relevant and well attended. It has longevity due to the legacy of what it achieves.

Scientist Susan Walker has presented at several EDS conferences: 'everyone turns up. It's a unique meeting of practitioners, policy makers, politicians and government bureaucrats. There is nothing else like it on the New Zealand scene. Presenting at the conferences has been a really important opportunity for me to put science in front of policy-makers.' This sentiment was echoed by Bryce Johnson, 'the EDS conference is the only conference in town really. The conferences have become absolutely critical and hugely important to the overall environmental debate in New Zealand.'

Fiona Driver grew up in Milford, Auckland, within a stone's throw of the Milford marina. Her father was a woodworking and technical drawing teacher and also a keen sailor. He built a 30-foot ketch in the family's front yard and finished his second boat off at the Milford marina. As a child, Driver spent many long school holidays sailing with her family around the Hauraki Gulf, which instilled in her a love of the outdoors. During her adult life, she has continued boating, and she currently owns, with her husband Rod Marler, a 46-foot Davidson sloop.

On leaving school, Driver initially studied chemistry, but during her OE was introduced to event management. On returning to New Zealand, she became a qualified professional conference organiser. She worked for The Conference Company in Auckland, then Arinex in Sydney. Driver eventually established her own event management consultancy called Driven Events.

COURTESY OF FIONA DRIVER

Driver first came into contact with EDS when she was contracted to work on its first Climate Change and Business Conference in 2004. In 2006 she joined EDS to manage the society's conferences and other events in-house. Driver went on to become general manager and one of the tight team of three (with Gary Taylor and Raewyn Peart) that ran EDS for a decade and a half. She continues to play a crucial role in running highly professional events and keeping EDS connected to the community.

During her time at EDS, a highlight for Driver has been 'seeing EDS and the causes we're fighting for becoming more "mainstream" and widely accepted and discussed. That was not how it was 15 plus years ago when a lot of the ideas were considered more "fringe".'

Driver left EDS at the end of 2020 to become Auckland branch manager of the New Zealand Law Society. She returned in early 2023, as 'I believe in what EDS stands for and enjoy feeling that I am contributing to making a meaningful difference to New Zealand and the world.' Fiona is now EDS's events director.

Epilogue: From the past to the future

AT THE TIME OF WRITING, New Zealand was still reeling from the impacts of Cyclone Gabrielle. The ferocious storm had devastated the North Island in early February 2023, just days after the Auckland Anniversary Day floods, when an atmospheric river of moisture-laden tropical air stalled over Auckland and dumped a month's worth of rain on the city in just one day. Billions of dollars of damage resulted and 11 people lost their lives. Scientists concluded that global warming had supercharged the weather events. The long-standing practice of building on floodplains is now being questioned and there is talk of managed retreat.

ALONG WITH THE WEATHER, much has changed in Aotearoa since EDS was first established in 1971. There were fewer than 3 million people living in the country then, a number that has increased to over 5 million today. Tāmaki Makaurau Auckland, the country's largest city, has almost tripled in size from a modest 650,000 residents in 1971 to over 1.7 million now. The city is bulging at the seams and a crisis of housing affordability and supply has prompted a drive for both urban intensification and sprawl throughout the Auckland region.

There have also been fundamental changes in the rural sector over the past five decades. The number of cows has doubled (from 3 to 6 million), while the number of sheep has halved. The rural economy has diversified and farming practices have intensified. The impacts have been most telling on the country's waterways. Close to half New Zealand's rivers are now categorised as 'unswimmable' due to the risk of bacterial infection. New Zealand's 'clean green' moniker is looking increasingly tattered.

The country's population is more multi-cultural and there has been a resurgence of Māori culture, language and economic might. Unlike in 1971, before the Waitangi Tribunal was established, most Treaty of Waitangi claims have now been settled. The Māori economy is dominant in primary production sectors, such as fishing, farming and forestry, and increasingly in urban development as well. The political power of Māori has also increased with Māori representation on some councils and Te Pati Māori (the Māori Party) having a growing presence in parliament.

Previous pages: Flooding at Piha after the Auckland Anniversary Day floods and Cyclone Gabrielle in 2023. RAEWYN PEART

At the same time, environmental awareness has widened and deepened amongst the general population. The state of freshwater became a significant issue in the 2017 general election and climate change featured in the top three issues of concern for voters in the run-up to the 2023 election. The younger generation has become more vocal on environmental issues, with the establishment of movements such as Extinction Rebellion and thousands of school children striking for action on the climate crisis.

Government has also significantly changed since EDS was first established. Instead of a Ministry *of Works and Development* we now have a Ministry *for the Environment*. Also, instead of the New Zealand Forest Service juggling forestry production and conservation functions, we now have a separate Department of Conservation to focus solely on the latter. There are fewer and better resourced councils, with the now-unified Auckland Council over-seeing an annual operating budget of some $5 billion. The important role that environmental groups play in holding the government and councils to account is now well accepted and their standing well entrenched in the law.

The laws overseeing the environment have also changed for the better. The Waters Pollution Act 1953, which oversaw egregious point-source pollution of the Waikato River, has been folded into the Resource Management Act and more recently the Natural and Built Environment Act. Resource extractive industries such as mining, which for a long time escaped any environmental scrutiny, have now been brought under robust environmental regulation.

The overall 'balancing' approach to decision-making, where economic considerations could trump environmental ones, is being replaced by environmental bottom lines. The King Salmon Supreme Court decision emphasised the importance of this focus. Te ao Māori concepts such as 'te mana o te wai', now enshrined in freshwater policy, emphasise the importance of ensuring the health of the natural environment before providing for its use. It's been a slow but positive transition.

Late last year a three-way coalition government between National, ACT and New Zealand First sought to wind back decades of progress on the natural environment by repealing the resource management reforms along with freshwater and biodiversity protections. There were also proposals to reinstate ministerial decision-making for development. Overall, this means that the role of EDS along with other environmental NGOs will become even more critical over coming years.

WHAT THEN DOES the future hold for Aotearoa New Zealand, and what role might EDS and other environmental organisations play over the next 50 years? There is no doubt that environmental challenges will become even more acute. The climate is changing and it is unclear to what extent the global community will be successful in limiting global warming to safe levels. There are increasing geopolitical uncertainties, with the Ukraine war causing significant political and economic disruptions and the increasing militarisation in the Pacific. At the same time, there is a growing biodiversity crisis, with tens of thousands of species threatened with extinction worldwide. The situation has become so dire that some have labelled it a sixth global mass extinction event, a situation unprecedented since the dinosaurs were wiped out some 65 million years ago.

The need for environmental defenders will likely only increase in the future as New Zealand faces the fundamental challenges of decarbonising the economy and addressing growing social inequities. Such challenges create both opportunities and risks. There is a risk that the nation's response to climate change will only place further pressure on the natural world, such as through extensive planting of permanent exotic carbon forests and proliferation of wind and solar farms over the country's remaining outstanding natural landscapes and significant biodiversity areas.

But the need to shift from business as usual also creates opportunities. There is the opportunity to create a different future, one which is more sustainable and where the restoration of nature becomes part of the solution. Urban areas could be 'greened' to reduce urban heating and improve the mental health of their inhabitants through better connection with nature, as well as provide new habitat for indigenous species.

Streams could be daylighted and rivers given room to roam to reduce flooding risks to settlements downstream. Marginal farmland could revert to healthy indigenous forests and marine habitats protected and restored to capture and store carbon and support a stable climate in the long term as well as thriving biodiversity.

EDS, along with others, will need to be in the thick of such critical choices facing New Zealand over coming decades, bringing a reasoned voice to the policy-making table on behalf of nature and holding decision-makers to account through the courts. It is hoped that through this chronicle of the past 50 years of EDS's activities, insights can be drawn to support and inspire the environmental defenders of the future.

EDS staff at the society's 2021 Christmas celebration. From left (top row) Sasha Maher, Cordelia Woodhouse and Greg Severinsen (middle row) Phoebe Parson, Deidre Koolen-Bourke, Tracey Turner and Shay Schlaepfer (front row) Bella Rollinson, Katharina Stickling, Raewyn Peart and Gary Taylor. DAVID HARTMAN

Appendix: EDS Publications and Conferences

Community Guide Series

Conservation in New Zealand: A Citizen's Guide to the Law. 1985. Alvin Smith and Gary Taylor.

The Community Guide to the Resource Management Act 1991. 2004. Raewyn Peart.

The Community Guide to Landscape Protection under the Resource Management Act 1991. 2005. Raewyn Peart.

Landscape Planning Guide for Peri-Urban and Rural Areas. 2005. Raewyn Peart.

The Community Guide to Coastal Development under the Resource Management Act 1991. 2005. Raewyn Peart.

The Community Guide to the Resource Management Act 1991 (2nd edition). 2007. Raewyn Peart.

The New Zealanders Guide to the Resource Management Act 1991 (3rd edition, published in association with Craig Potton Publishing). 2008. Raewyn Peart.

Managing Freshwater. 2010. Raewyn Peart, Kate Mulcahy and Natasha Garvan.

Strengthening Second Generation Regional Policy Statements. 2011. Raewyn Peart and Peter Reaburn.

Caring for Our Coast: An EDS Guide to Managing Coastal Development. 2013. Lucy Brake and Raewyn Peart (winner of the 2013 Resource Management Law Association Publication Award).

Treasuring our Biodiversity: An EDS Guide to the Protection of New Zealand's Indigenous Habitats and Species. 2013. Lucy Brake and Raewyn Peart.

Sustainable Seas: Managing New Zealand's Marine Environment. 2015. Lucy Brake and Raewyn Peart.

Web-based Environment Guide at www.environmentguide.org.nz

Environmental Policy Conferences

The New Zealand Landscape Conference. 25–26 July 2003. Auckland.

New Zealand Coastal Conference: Development and Conservation of our Coasts and Lakesides. 2004. Auckland.

Seachange 05: Managing Our Coastal Waters and Oceans. 21–22 November 2005. Auckland.

Beyond the RMA: An In-depth Exploration of the Resource Management Act 1991. 30–31 May 2007. Auckland.

Conflict in Paradise: The Transformation of Rural New Zealand. 11–12 June 2008. Auckland.

Reform in Paradise: Threat or Opportunity. 8–9 June 2009. Auckland.

Reform in Paradise II. 2–3 June 2010. Auckland.

Coastlines: Spatial Planning for Land and Sea. 2 June 2011. Auckland.

Pathways to Paradise? Farming, Forestry and Mining. 6–7 August 2012. Auckland.

Our Place: State of the Environment. 8 August 2013. Auckland.

Navigating Our Future: Addressing Risk and Building Resilience. 6–7 August 2014. Auckland.

Wild Things: Addressing Terrestrial, Freshwater and Marine Biodiversity Loss. 12–13 August 2015. Auckland.

Wild Places. 10–11 August 2016. Auckland.

Tipping Points. 9–10 August 2017. Auckland.

Green Light or Light Green?: The Government's Environmental Reforms. 1–2 August 2018. Auckland.

Through New Eyes: Rethinking Landscape in Aotearoa. 14–15 August 2019. Auckland.

Transforming Aotearoa: The Government's Environmental Reform Agenda. 4–6 August 2021. Christchurch.

Pivot Point: Deep Environmental Change. 22–24 March 2023. Auckland.

Climate Change and Business Conferences

1st Australia-New Zealand Climate Change and Business Conference. November 2004. Auckland.

2nd Australia-New Zealand Climate Change and Business Conference. February 2006. Adelaide.

3rd Australia-New Zealand Climate Change and Business Conference. 29–31 August 2007. Brisbane.

4th Australia-New Zealand Climate Change and Business Conference. 18–20 August 2008. Auckland

5th Australia-New Zealand Climate Change and Business Conference. 24–25 August 2009. Melbourne.

6th Australia-New Zealand Climate Change and Business Conference. 10–12 August 2010. Sydney.

7th Australia-New Zealand Climate Change and Business Conference. 1–2 August 2011. Wellington.

8th New Zealand Climate Change and Business Conference. 20–21 October 2015. Auckland.

9th New Zealand Climate Change and Business Conference. 11–12 October 2016. Auckland.

10th New Zealand Climate Change and Business Conference. 10–11 October 2017. Auckland.

11th New Zealand Climate Change and Business Conference. 9–10 October 2018. Auckland.

12th New Zealand Climate Change and Business Conference. 8–9 October 2019. Auckland.

13th New Zealand Climate Change and Business Conference. 6–7 October 2020. Auckland.

14th New Zealand Climate Change and Business Conference. 19–20 September 2022. Auckland.

15th New Zealand Climate Change and Business Conference. 19–20 September 2023. Auckland.

Policy Reports

Landscape

Reclaiming Our Heritage: The Proceedings of the New Zealand Landscape Conference, 25-26 July 2003. 2003. Environmental Defence Society.

A Place to Stand: The Protection of New Zealand's Natural and Cultural Landscapes. 2004. Raewyn Peart.

Castles in the Sand: What's Happening to the New Zealand Coast? (in association with Craig Potton Publishing). 2009. Raewyn Peart.

Tourism and Landscape Protection. 2020. Raewyn Peart and Cordelia Woodhouse.

Te Manahuna Mackenzie Basin and Landscape Protection. 2020. Raewyn Peart and Cordelia Woodhouse.

Protecting the Waitākere Ranges. 2020. Cordelia Woodhouse, Shay Schlaepfer and Raewyn Peart.

Protecting the Hauraki Gulf Islands. 2020. Raewyn Peart and Cordelia Woodhouse.

Caring for the Landscapes of Aotearoa New Zealand: Synthesis Report. 2021. Raewyn Peart, Cordelia Woodhouse, Shay Schlaepfer, Deidre Koolen-Bourke and Lara Taylor.

Restoring Te Pātaka o Rākaihautū Banks Peninsula. 2021. Raewyn Peart and Cordelia Woodhouse.

Marine and freshwater

Looking Out to Sea: New Zealand as a Model for Oceans Governance. 2005. Raewyn Peart.

Governing the Oceans: Environmental Reform for the Exclusive Economic Zone. 2011. Raewyn Peart, Kelsey Serjeant and Kate Mulcahy.

Beyond the Tide: Integrating the Management of New Zealand's Coast (1st and 2nd editions). 2011. Raewyn Peart, Kelsey Serjeant and Kate Mulcahy.

Safeguarding Our Oceans: Strengthening Marine Protection in New Zealand. 2012. Kate Mulcahy, Raewyn Peart and Abbie Bull.

Wonders of the Sea: The Protection of New Zealand's Marine Mammals. 2012. Kate Mulcahy an
d Raewyn Peart.

Dolphins of Aotearoa: Living with Dolphins in New Zealand (in association with Craig Potton Publishing; shortlisted for the 2015 Royal Society of New Zealand Science Book Prize). 2013. Raewyn Peart.

The Story of the Hauraki Gulf: Discovery, Transformation, Restoration (in association with David Bateman Limited). 2016. Raewyn Peart

Turning the Tide: Integrated Marine Planning in New Zealand. 2018. Raewyn Peart.

Governance of the Hauraki Gulf: A Review of Options. 2018. Raewyn Peart and Brooke Cox.

Voices from the Sea: Managing New Zealand's Fisheries. 2018. Raewyn Peart.

Farming the Sea: Marine Aquaculture with Resource Management System Reform. 2019. Raewyn Peart.

Healthy Seas: Implementing Marine Spatial Planning in New Zealand. 2019. Kelsey Serjeant and Raewyn Peart.

The Breaking Wave: A Conversation About Reforming the Oceans Management System in Aotearoa New Zealand. 2021. Greg Severinsen, Raewyn Peart and Bella Rollinson.

The Breaking Wave: Oceans Reform in Aotearoa New Zealand. 2022. Greg Severinsen, Raewyn Peart, Bella Rollinson, Tracey Turner and Phoebe Parson.

Science for Policy: The Role of Science in the National Policy Statement for Freshwater Management. 2022. Deidre Koolen-Bourke and Raewyn Peart.

Resource management

Beyond the RMA: An In-depth Exploration of the Resource Management Act 1991. Conference proceedings. 2007. Environmental Defence Society.

Improving Environmental Governance: The Role of an Environmental Protection Authority in New Zealand. 2009. Raewyn Peart.

Report on the Review of Sections 6 and 7 of the Resource Management Act 1991. 2012. Environmental Defence Society Technical Advisory Group

Evaluating the Environmental Outcomes of the RMA (in association with Infrastructure New Zealand, Employers and Manufacturers Association and Property Council of New Zealand). 2016. Marie Brown, Raewyn Peart and Madeleine Wright.

Last Line of Defence: Compliance, Monitoring and Enforcement of New Zealand's Environmental Law. 2017. Marie Brown.

Reform of the Resource Management System: The Next Generation, working paper 1. 2018. Greg Severinsen and Raewyn Peart.

Reform of the Resource Management System: The Next Generation, working paper 2. 2018. Greg Severinsen, Raewyn Peart and Brooke Cox.

Reform of the Resource Management System: The Next Generation, working paper 3. 2018. Greg Severinsen and Raewyn Peart.

Reform of the Resource Management System: The Next Generation, Synthesis Report (winner of the 2019 Resource Management Law Association Publication Award). 2018. Greg Severinsen and Raewyn Peart.

Reform of the Resource Management System: A Pathway to Reform, Working Paper 1: Criteria for Reform. 2019. Greg Severinsen.

Reform of the Resource Management System: A Pathway to Reform, Working Paper 2: A Model for the Future. 2019. Greg Severinsen.

Reform of the Resource Management System: A Model for the Future, Synthesis Report. 2019. Greg Severinsen.

A Review of the Resource Management (National Environmental Standards for Plantation Forestry) Regulations 2017. 2019. Madeleine Wright, Sally Gepp and David Hall.

Reform of the Resource Management System: The Urban Context. 2020. Greg Severinsen.

The Strategic Planning Act and Funding. 2021. Greg Severinsen and Raewyn Peart.

Designing the Strategic Planning Act to Better Address the Marine Environment. 2021. Raewyn Peart and Greg Severinsen.

Landscape Protection in a Future Resource Management System. 2021. Raewyn Peart and Cordelia Woodhouse.

National Direction in a Future Resource Management System. 2021. Greg Severinsen and Raewyn Peart.

Environmental Limits in a Future Resource Management System. 2021. Greg Severinsen and Raewyn Peart.

Environmental Advocacy in the Future Resource Management System. 2023. Greg Severinsen.

Aotearoa New Zealand's Climate Change Adaptation Act: Building a Durable Future, Principles and Funding for Managed Retreat, Working Paper 1. 2023. Raewyn Peart, Jonathan Boston, Sasha Maher and Teresa Konlechner.

Aotearoa New Zealand's Climate Change Adaptation Act: Building a Durable Future, Current Legislative and Policy Framework for Managed Relocation, Working Paper 2. 2023. Raewyn Peart and Benjamin Tombs.

Biodiversity/Conservation

Vanishing Nature: Facing New Zealand's Biodiversity Crisis. 2015. Marie Brown, Theo Stephens, Raewyn Peart and Bevis Fedder.

Pathways to Prosperity: Safeguarding Biodiversity in Development. 2016. Marie Brown.

Banking on Biodiversity: The Feasibility of Biodiversity Banking in New Zealand. 2017. Marie Brown.

Conserving Nature: Conservation Reform Issues Paper (winner of the 2021 Resource Management Law Association Publication Award). 2021. Deidre Koolen-Bourke and Raewyn Peart.

Independent Review of the Conservation Management Planning System. 2023. Deidre Koolen-Bourke, Raewyn Peart, Bronwyn Wilde and Tracey Turner.

Reform of the Wildlife Act 1953: An Opportunity for Transformational Change of Aotearoa New Zealand's Biodiversity Law. 2023. Deidre Koolen-Bourke, Raewyn Peart and Shay Schlaepfer.

Independent Review of the Conservation Management Planning System. 2023. Deidre Koolen-Bourke, Raewyn Peart, Bronwyn Wilde and Tracey Turner.

References and court decisions

Chapter 1 Environmental defender needed

Court decisions (in order mentioned)

Huntly Borough v Williams [1974] 1 NZLR 689 (Court of Appeal decision on the Huntly raw sewage discharge)

Environmental Defence Society Inc v Agricultural Chemicals Board [1973] 2 NZLR 758 (Supreme Court decision on the use of 2,4,5-T)

R. T. Mahuta v National Water and Soil Conservation Authority [1973] A571 (Town and Country Appeal Board decision on the Huntly Power Station water rights)

References

Anon. 1973. Writ sought in Supreme Court for action on herbicide 2,4,5-T. *The Post*, 16 April.

Anon. 1973. 2,4,5-T concern 'heightened'. *The Dominion*, 17 April.

Environmental Defence Society. 1972. The case against 2,4,5-T. *New Zealand Environment* 2(2): 16–21.

Environmental Defence Society. 1974. Herbicide 2,4,5-T – Agricultural Chemicals Board fails to protect New Zealanders. Press release, 14 September.

Fookes, T. W. 1978. Social and economic impact of the Huntly Power Station, New Zealand. *Ekistics* 45(26): 200–6.

Fowles, J., L. Gallagher, V. Baker, D. Phillips, F. Marriot, C. Stevenson and M. Noonan. 2005. *A Study of 2,3,7,8-Tetrachlorodibenzo-p-dioxin (TCDD) Exposures in Paritutu, New Zealand*. A report to the Ministry of Health.

Mann, B. 1972. 2,4,5-T: an unregulated danger. *Soil and Health Journal* April-May: 20–2.

McGregor, J. 1972. Pollution charge hearing due soon. *The Dominion*. 17 July.

McMahon, C. K. (ed). 1972. *The Physical Environment Conference 1970: Reports, Papers and Proceedings*. Wellington: Environmental Council. 1972.

Sare, W. M. and P. I. Forbes. 1971. Letter on deformed babies at Te Awamutu. *New Zealand Medical Journal* Vol 75, 476: 37–8.

Statistics New Zealand. 1970. *The New Zealand Official Yearbook*. Wellington: Statistics New Zealand digital yearbook collection.

Whittle, J. 2013. Into the backyard: Huntly power station and the history of environmentalism in New Zealand. *Environment and Nature in New Zealand* 8(1): 1.

Williams, D. 2011. Speech for the Environmental Defence Society at its 40-year anniversary celebration.

Wurtser, C. G. 2015. *DDT Wars: Rescuing Our National Bird, Preventing Cancer, and Creating the Environmental Defense Fund*. New York: Oxford University Press.

Chapter 2 Power play

Court decisions (in order mentioned)

Liquigas Limited v The Manukau City Council [1983] NZPT 79 (Planning Tribunal decision on application for LPG storage facility at Ash Road)

Environmental Defence Society v Manukau City Council [1986] NZPT 31 (Planning Tribunal decision on application for LPG storage facility at McLaughins Road)

References

Anon. 1976. Proposed power plant 'rape of Maori Land'. *New Zealand Herald*, 16 February.

Anon. 1977. Scientist: Don't tell public nuclear faults. *Auckland Star*, 27 February.

Anon. 1977. Waiau Pa farce. *Auckland Star*, 5 July.

Bioresearches Limited. 1976. *The Ecological Impact of the Proposed Development of Auckland Thermal Number One at Waiau Pa*. Auckland: A report prepared for the New Zealand Electricity Department.

Energy in New Zealand: Proceedings of the New Zealand Energy Conference 1974. University of Auckland, 23–25 May.

Herald energy reporter. 1976. Waiau Pa site objectors in losing battle. *New Zealand Herald*, 20 December.

Lewis, O. and J. Lee. 2018. From uranium to nuclear plants: New Zealand's 'secret' nuclear past. *Sunday Star Times*, 14 January.

McLarin, N. 1978. In memory of Auckland Thermal No. 1 Power Station, Waiau Pa. *Town Planning Quarterly* 53: 9-11.

Ministry of Works and Development. 1976. Minutes of the 1st Meeting on Auckland Thermal No 1 Power Station. Friday, 23 July.

New Zealand Electricity Department. 1971. Proposed thermal power station in Auckland Area. Minutes of Second Liaison Committee meeting. 15 January.

New Zealand Electricity Department. 1974. *Preliminary environmental impact report: Auckland Thermal No. 1 Power Station*. Wellington.

Royal Commission of Inquiry into Nuclear Power. 1978. *Nuclear Power Generation in New Zealand.*

Tattersfield, B. 2020. *Bill Birch: Minister of Everything.* Mangawhai: Mary Egan Publishing.

Waitangi Tribunal. 1978. Report of the Waitangi Tribunal on the Waiau Power Station Claim (Wai 2). Wellington: Department of Justice.

Wilson, R. 1982. *From Manapouri to Aramoana: The Battle for New Zealand's Environment.* Auckland: Earthworks Press.

Chapter 3 Thinking bigger

Court decisions (in order mentioned)

Environmental Defence Society v South Pacific Aluminium Limited [1981] 1 NZLR 146 (Court of Appeal interim decision on discovery regarding the Aramoana smelter application)

Environmental Defence Society v South Pacific Aluminium Limited (No 2) [1981] 1 NZLR 153 (Court of Appeal final decision on discovery regarding the Aramoana smelter application)

Environmental Defence Society v South Pacific Aluminium Limited (No 3) [1981] 1 NZLR 216 (Court of Appeal decision on bringing the Aramoana smelter application under the National Development Act)

Environmental Defence Society v South Pacific Aluminium Limited (No 4) [1981] 1 NZLR 530 (Court of Appeal decision on environmental impact report for Aramoana smelter)

Environmental Defence Society v National Water and Soil Conservation Authority [1979] 7 NZPT 385 (judicial review of Water and Soil Conservation Authority decision to grant Clyde Dam water rights)

McGregor v Attorney-General A 16/78 (application to stop work at Clyde Dam)

Annan v National Water and Soil Conversation Authority and Minister of Energy [1982] 7 NZPT 417 (Planning Tribunal decision to grant Clyde Dam water rights)

Gilmore v National Water and Soil Conservation Authority and Minister of Energy [1982] 8 NZTPA 298 (High Court decision to overturn grant of Clyde Dam water rights)

Annan v National Water and Soil Conversation Authority and Minister of Energy [1982] 8 NZTPA 369 (Planning Tribunal's reconsideration and decline of the Clyde Dam water rights)

References

Hansard debates. 1982.

McDowell, R. M. 1994. *Gamekeepers for the Nation: The Story of New Zealand's Acclimatisation Societies 1861–1990.* Christchurch: Canterbury University Press.

Ministerial Review Committee. 1990. *Clyde Dam Project Decision-making.* Report to Cabinet, Wellington.

Nathan, S. 2021. Geology and the Clyde Dam. *NZ Geomechanics News* 101, June: 42–52.

Powell, P. 1978. *Who Killed the Clutha?* Dunedin: John McIndoe.

Tattersfield, B. 2020. *Bill Birch: Minister of Everything.* Mangawhai: Mary Egan Publishing.

Wilson, R. 1982. *From Manapouri to Aramoana: The Battle for New Zealand's Environment.* Auckland: Earthworks Press.

Chapter 4 Wild and scenic

Court decisions (in the order mentioned)

In the matter of National Water Conservation (Motu River) Order 1983, NZPT, W5/84 (Planning Tribunal decision on the Motu River Water Conservation Order)

Ashburton Acclimatisation Society v Federated Farmers of New Zealand Inc [1988] 1 NZLR 78 (Court of Appeal decision on the Rakaia River Water Conservation Order)

References

Commission for the Environment. 1977. *Wild and Scenic Rivers Protection: A Discussion Paper.* Wellington: Office of the Minister for the Environment.

Commission for the Environment. 1978. *Wild and Scenic Rivers Protection: An Appraisal.* Wellington: Commission for the Environment.

MacMurray, C. 1984. *The Motu Dialogue.* Report prepared for the Recreation and Landscape Values Working Party.

Oldham, C. D. C. 1989. 'Wild and Scenic River Conservation in New Zealand'. Masters thesis, University of Canterbury.

Penny, S. 1982. *Motu: A Wild and Scenic River.* A review prepared for the Environmental Defence Society. Auckland: EDS.

Tattersfield, B. 2020. *Bill Birch: Minister of Everything.* Mangawhai: Mary Egan Publishing.

Williams, D. 2023. Court puts river protection case on fast-track. Newsroom, 27 February 2023.

Young, D. 2013. *Rivers: New Zealand's Shared Legacy.* Auckland: Random House.

Chapter 5 Into hot water

Court decisions (in order mentioned)

Keam v National Water and Soil Conservation Authority [1979] NZPT 202 (Planning Tribunal decision declining the Minister of Works and Development Waimangu application)

Minister of Works and Development v Keam [1981] 7 NZTPA 289 (High Court decision overturning Planning Tribunal decision on the Waimangu application)

Keam v Minister of Works and Development [1982] 1 NZLR 319 (Court of Appeal decision reinstating Planning Tribunal decision declining the Waimangu application)

Minister of Lands v Bay of Plenty Regional Water Board and F J Ramsey Limited v Bay of Plenty Regional Water Board [1984] 10 NZTPA 161 (Planning Tribunal decision on the F. J. Ramsey application)

References

Anon. 1982. Massive claims could follow bore closures. *Dominion Post*, 2 August.

Anon. 1982. Newspaper accused of alarmist reports. *Dominion Post*, 5 August.

Anon. 1982. Geothermal draw-off not affecting geysers or springs. *Dominion Post*, 12 August.

Anon. 1983. Task force wanted to save Whaka. *New Zealand Herald*, 25 May

Gordon, D. A., B. J. Scott and E. K. Mroczek. 2005. *Rotorua Geothermal Field Management Monitoring Update: 2005*. Whakatāne: Environmental Bay of Plenty.

Houghton, B. F., E. F. Lloyd and R. F. Keam. 1980. *The Preservation of Hydrothermal System Features of Scientific and Other Interest*. A report for the Geological Society of New Zealand. Wellington: Natural Conservation Council.

Keam, R. F. 1982. 'Rotorua Geothermal System: Comments on Geyser Preservation and on System Stability'. Unpublished paper.

Lloyd, E. F. Undated. 'Report on the April 1983 Pressure Drop at Ororea Springs, Te Roto-A-Tamaheke, Whakarewarewa'. Unpublished paper.

Report of the Rotorua Geothermal Task Force, July 1983.

Scott, B. J. and A. D. Cody. Undated. 'Effects of Bore Closure at Rotorua, New Zealand'. Unpublished paper.

Scott, B. J., D. A. Gordon and A. D. Cody. 2005. Recovery of Rotorua geothermal field, New Zealand: Progress, issues and consequences. *Geothermics* 34: 161–185.

Scott, B. 2019. *Rotorua Geothermal System: The Science Story Environmental Summary Report*. Bay of Plenty Regional Council.

Chapter 6 Digging deeper

Court decisions (in order mentioned)

Environmental Defence Society Inc v Patterson [1981] 2 NZLR 754 (High Court decision on the Gold Mines NZ Kūaotunu prospecting case)

Environmental Defence Society v The Planning Tribunal [1987] M697/87 (High Court decision on Waiomu mine licence conditions)

Application by Spectrum Resources Ltd for a mining licence [1987] W47/87 (Decision on adjournment application on hearing the mining licence for the Monowai mine)

Spectrum Resources Ltd v Minister of Conservation [1989] 3 NZLR 351 (High Court decision on Monowai mine land access decision)

References

Gooding, B. 1982. Coromandel: The Peninsula war. *New Zealand Listener*, 4 December: 32–4.

Tanber, G. J. 1993. Almost paradise. *Chicago Tribune*, 1 January.

Wilson, R. 1982. *From Manapouri to Aramoana: The Battle for New Zealand's Environment*. Auckland: Earthworks Press.

Chapter 7 Protecting the coast

Court decisions (in order mentioned)

R. A. Houston and Environmental Defence Society v Thames County Council [1973] 4 NZTPA 9 (Town and Country Planning Appeal Board decision approving development at Onemana)

The Physical Environment Association of Coromandel Inc v Thames-Coromandel District Council [1977] 782/75 (Town and Country Planning Appeal Board decision declining structure plan for Matarangi)

The Physical Environment Association of Coromandel Inc v Thames-Coromandel District Council [1982] A70/82 (Town and Country Planning Appeal Board decision declining development of the Hahei south headland)

Environmental Defence Society v Great Barrier Island County Council [1977] 6 NZTPA 301 (Town and Country Planning Appeal Board declining variation to Great Barrier Island District Scheme)

H. Burkhardt v Mangonui County Council [1979] NZPT 185 (first Planning Tribunal decision removing any reference to the proposed Karikari development in the district scheme)

Environmental Defence Society v Mangonui County Council [1986] NZPT 13 (second Planning Tribunal decision authorising stage 1 of the Karikari development)

Environmental Defence Society v Mangonui County Council [1987] 12 NZTPA 349 (High Court decision upholding Planning Tribunal decision on Karikari development)

Environmental Defence Society v Mangonui County Council [1989] 13 NZTPA 197 (Court of Appeal decision overturning Planning Tribunal decision on Karikari development)

Opoutere Residents and Ratepayers Association v Thames-Coromandel District Council [1985] NZTPA 246 (Planning Tribunal interim decision granting consent to a reduced camping ground development at Opoutere)

Opoutere Residents and Ratepayers Association v The Planning Tribunal [1989] 13 NZTPA 446 (Court of Appeal decision declining consent for the proposed camping ground at Opoutere)

References

Anon. 1971. Luxury resort plan for East Coast. *The Dominion*, 25 September.

Anon. 1971. Waikawau plans now 'like assault causing murder'. *The Waikato Times*, 28 October.

Furey, L. 1991. Excavations at Whitipirorua, T12/16, Coromandel Peninsula. *Records of the Auckland Institute and Museum* 28: 1–32.

Mutu, M. 2010. 'Ngai Kahu Kaitiakitanga' in R. Selby, P. Moore and M. Mulholland (eds), *Māori and the Environment: Kaitiaki*. Wellington: Huia.

Peart, R. 2009. *Castles in the Sand: What's Happening to the New Zealand Coast?* Nelson: Craig Potton Publishing.

Peart, R. 2016. *The Story of the Hauraki Gulf: Discovery, Transformation and Restoration*. Auckland: David Bateman Limited in association with the Environmental Defence Society.

Chapter 8 End of an era

References

Severinsen, G. 2023. *Environmental Advocacy in the Future Resource Management System*. Auckland: Environmental Defence Society.

Chapter 9 EDS reborn

Court decisions (in order mentioned)

Arrigato Investments Limited v Rodney District Council [1999] NZEnvC 399 (Environment Court interim decision on development on Pakiri south headland)

Bayley v Manukau City Council [1999] 1 NZLR 658 (Court of Appeal decision establishing the permitted baseline test)

Far North District Council v Te Runanga-A-Iwi O Ngati Kahu [2013] NZCA 221 (Court of Appeal decision that further development of the Karikari property was not in breach of agreement with Ngāti Kahu and EDS)

Coromandel Watchdog of Hauraki Inc v Chief Executive of the Ministry of Economic Development [2007] NZCA 473 (Court of Appeal decision allowing use of prohibited activity status for mining in the Thames-Coromandel District Plan)

Environmental Defence Society v Taranaki Regional Council [2002] NZEnvC 441 (Environment Court decision approving the Stratford combined cycle gas-fired power plant)

Environmental Defence Society v Auckland Regional Council [2002] NZRMA 492 (Environment Court decision approving the Ōtāhuhu combined cycle gas-fired power plant)

References

Leining, C. 2022. *A Guide to the New Zealand Emissions Trading Scheme: 2022 update*. Wellington: Motu Economic and Public Policy Research.

Chapter 10 Protecting the coast ... again

Court decisions (in order mentioned)

Tairua Environment Society v Thames-Coromandel District Council NZEnvC A97/2004 (Environment Court Decision on Hot Water Beach esplanade reserve)

Nga Uri O Wiremu Mormona Rau Ko Whakarongohau Pita Incorporated v Far North District Council [2008] NZEnvC 37 (interim decision of the Environment Court on the proposed Ngaiotonga coastal development)

Northern Land Property Limited v Thames-Coromandel District Council [2021] NZEnvC 180 (interim decision of the Environment Court on a structure plan for development of land behind New Chums Beach)

References

Anyan, S. 2011. Showdown at New Chums. *North and South*, January.

Barrington, M. 2012. Old investment charges, lower penalty. *Northern Advocate*, 2 May.

Beston, A. 2003. Buyers warned off university's land. *New Zealand Herald*, 16 January.

Beston, A. 2003. PM asks varsity to rethink coastal land sale. *New Zealand Herald*. 10 February.

Dinsdale, M. 2021. Campaign to buy land next to Northland's Ngunguru sandspit and return it to public ownership. *Northern Advocate*. 13 December.

Dinsdale, M. 2022. Plan to buy land at Northland's Ngunguru sandspit for public ownership over due to lack of funds. *Northern Advocate*. 24 March.

Gibbs, J. G. 1977. Late Quaternary sedimentary processes at Ohiwa Harbour eastern Bay of Plenty with special reference to property loss on Ohiwa Spit. *Water and Soil Technical Publication* No 5.

Hart, R. 2014. *Assessment of Landscape and Visual Effects for a Proposed Subdivision at Wainuiototo Farm*. Rotorua: Wildlands.

Peart, R. 2009. *Castles in the Sand: What's Happening to the New Zealand Coast?* Nelson: Craig Potton Publishing.

Piper, D. 2022, Push for hapū ownership of Ngunguru sandspit and maunga to displace public bid. *Stuff*, 14 February.

Turbott, C. 2006. *Managed Retreat from Coastal Hazards: Options for Implementation*. Hamilton: Environment Waikato.

Chapter 11 Natural landscapes under threat

Court decisions (in order mentioned)

Environmental Defence Society v Kaipara District Council [2010] NZEnvC 294 (Environment Court decision requiring the Kaipara District Council to address landscape issues in its district plan – the 'blank pages' case)

Environmental Defence Society v Taupo District Council [2009] NZEnvC 235 (Environment Court decision on Plan Change 19 which brought in stronger controls on rural subdivision in the Taupō District Plan)

Mapara Valley Preservation Society v Taupo District Council [2009] NZEnvC 233 (Environment Court decision turning down the Lakeview Ventures development in the Mapara Valley)

Sade Developments No 2 Ltd v Taupo District Council [2009] NZEnvC 234 (Environment Court decision turning down the proposed development of Whakaroa Point in Taupō)

References

Environmental Defence Society. 2003. *Reclaiming Our Heritage: The Proceedings of the New Zealand Landscape Conference*. 25–26 July 2003. Auckland: Environmental Defence Society.

Peart, R. 2004. *A Place to Stand: The Protection of New Zealand's Natural and Cultural Landscapes*. Auckland: Environmental Defence Society.

Chapter 12 Protecting the Mackenzie

Court decisions (in order mentioned)

High Country Rosehip Orchards v Mackenzie District Council [2011] NZEnvC 387 (Environment Court's first interim decision on Plan Change 13 to the Mackenzie District Plans)

Environmental Defence Society v Mackenzie District Council [2016] NZEnvC 253 (Environment Court decision on a declaration to close vegetation clearance loopholes in the Mackenzie District Plan)

Federated Farmers v Mackenzie District Council [2017] NZEnvC 53 (Environment Court's eleventh decision on Plan Change 13 which confirmed its amended provisions)

Mackenzie District Council [2019] NZEnvC 56 (Environment Court declaration on activity status for Simons Pass Station application)

Simons Pass Station Ltd v Mackenzie District Council [2020] NZHC 3265 (High Court declaration on activity status for Simons Pass Station application)

References

Peart, R. and C. Woodhouse. 2020. *Te Manahuna-Mackenzie Basin and Landscape Protection*. Auckland: Environmental Defence Society.

Peart, R., C. Woodhouse, S. Schlaepfer, D. Koolen-Bourke and L. Taylor. 2021. *Caring for the Landscapes of Aotearoa New Zealand: Synthesis Report*. Auckland: Environmental Defence Society.

Chapter 13 Cleaning up rivers

Court decisions (in order mentioned)

Applications by Central Plains Water, joint decisions and recommendation of Independent Commissioners, 28 May 2010 (Decision of the Independent Commissioners approving the Central Plains Water scheme but without the dam)

Final report and decisions of the Board of Inquiry into the Tukituki Catchment Proposal, 18 June 2014 (Board of Inquiry decision on the plan change to facilitate the Ruataniwha dam imposing a dual-nutrient approach)

Hawke's Bay and Eastern Fish and Game Councils v Hawke's Bay Regional Council [2014] NZHC 3191 (High Court decision overturning parts of the Board of Inquiry's decision on the Ruataniwha dam-related plan change)

Wellington Fish and Game Council and Environmental Defence Society v Manawatu-Wanganui Regional Council [2017] NZEnC 37 (High Court decision on the legality of the council's implementation of the Horizons One Plan)

References

Hutching, G. 2014. English pushes case for Ruataniwha dam. *Stuff*, 2 July.

Joy, M. 2021. Vested interests in big agriculture: A freshwater scientist's personal experience. *Policy Quarterly*, 17(2): 51–5.

Koolen-Bourke, D. and R. Peart. 2022. *Better Linking Science to Policy: An Examination of the Role of Science in the Development of the National Policy Statement for Freshwater Management*. Auckland: Environmental Defence Society.

Oram R. 2009. Forum suggestion holds water. *Stuff*, 17 July.

Oram, R. 2010. Water forum offers sign of hope. *Stuff*, 26 September.

Chapter 14 Better managing the sea

Court decision

Akaroa Marine Protection Society Inc v Minister of Conservation [2012] NZRMA 343 (High Court decision overturning the Minister's decision to decline the Akaroa Marine Reserve)

References

Hauraki Gulf Forum. 2011. *Spatial Planning for the Gulf: An International Review of Marine Spatial Planning Initiatives and Application to the Hauraki Gulf*. Written for the Forum by Raewyn Peart. Auckland: Hauraki Gulf Forum.

Milne, J. 2020. Winston Peters puts his mouth where his money is. *Newsroom*, 30 September.

Mulcahy, K. and R. Peart. 2012. *Wonders of the Sea: The Protection of New Zealand's Marine Mammals*. Auckland: Environmental Defence Society.

Mulcahy, K., R. Peart and A. Bull. 2012. *Safeguarding Our Oceans: Strengthening Marine Protection in New Zealand*. Auckland: Environmental Defence Society.

Peart, R. 2005. *Looking Out to Sea: New Zealand as a Model for Oceans Governance*. Auckland: Environmental Defence Society.

Peart, R. 2013. *Dolphins of Aotearoa: Living with Dolphins in New Zealand*. Nelson: Craig Potton Publishing in association with the Environmental Defence Society.

Peart, R. 2016. *The Story of the Hauraki Gulf: Discovery, Transformation, Restoration*. Auckland: David Bateman Limited in association with the Environmental Defence Society.

Peart, R. 2018. *Voices From the Sea: Managing New Zealand's Fisheries*. Auckland: Environmental Defence Society.

Peart, R. 2019. *Farming the Sea: Marine Aquaculture with Resource Management System Reform*. Auckland: Environmental Defence Society.

Peart, R., K. Serjeant and K. Mulcahy. 2011. *Governing the Oceans: Environmental Reform for the Exclusive Economic Zone*. Auckland: Environmental Defence Society.

Serjeant, K. and R. Peart. 2019. *Healthy Seas: Implementing Marine Spatial Planning in New Zealand*. Auckland: Environmental Defence Society.

Severinsen, G., R. Peart, B. Rollinson, T. Turner and P. Parsons. 2022. *The Breaking Wave: Oceans Reform in Aotearoa New Zealand*. Auckland: Environmental Defence Society.

Chapter 15 Strengthening the system

Court decisions (in order mentioned)

Environmental Defence Society Inc v New Zealand King Salmon Company Ltd [2013] NZHC 1992 (High Court decision confirming New Zealand King Salmon applications)

Environmental Defence Society v New Zealand King Salmon Company Ltd [2014] NZSC 38 (Supreme Court decision turning down New Zealand King Salmon Port Gore application)

Man O'War Station Ltd v Auckland Council [2015] NZHC 767 (High Court decision on the mapping of outstanding natural landscapes)

Man O' War Station Ltd v Auckland Council [2017] NZCA 24 (Court of Appeal decision on mapping of outstanding natural landscapes)

Man O'War Station Ltd v Auckland Council [2017] NZHC 3217 (High Court decision on rural activities in outstanding natural landscapes)

Environmental Defence Society v Otago Regional Council [2019] NZHC 2278 (High Court decision on requirement for ports to adhere to avoidance policies)

Port Otago Limited v Environmental Defence Society [2021] NZCA 638 (Court of Appeal decision on requirement for ports to adhere to avoidance policies)

Port Otago Limited v Environmental Defence Society [2023] NZSC 112 (Supreme Court decision on the requirement for ports to adhere to avoidance policies)

References

Anon. 2020. Handshake deal to clean up dross after Mataura floods goes sour. *Radio New Zealand*, 17 February.

Brown, M. 2017. *Last Line of Defence: Compliance Monitoring and Enforcement of New Zealand's Environmental Law*. Auckland: Environmental Defence Society.

Brown, B., R. Peart and M. Wright. 2016. *Evaluating the Environmental Outcomes of the RMA*. Auckland: Environmental Defence Society in association with Infrastructure New Zealand, Employers and Manufacturers Association and Property Council of New Zealand.

Brown, B., T. Stephens, R. Peart and B. Fedder. 2015. *Vanishing Nature: Facing New Zealand's Biodiversity Crisis*. Auckland: Environmental Defence Society.

Environmental Defence Society v New Zealand Aluminium Smelters Limited ENV 200-CHC, First affidavit of Shay Schlaepfer on behalf of Environmental Defence Society Inc in support of application for declarations, 3 July 2020.

Houlahan, M. 2020. Toxic gas fear at ouvea site. *Otago Daily Times*, 20 July.

Jackson, B. 2021. Rio Tinto to pick up $6 million ouvea pre-mix clean-up bill. *Stuff*, 27 May.

Koolen-Bourke, D and R. Peart. 2021. *Conserving Nature: Conservation Reform Issues Paper*. Auckland: Environmental Defence Society.

New Zealand Aluminium Smelter. 2020. Media statement, 17 February.

Peart, R. 2009. *Improving Environmental Governance: The Role of an Environmental Protection Authority in New Zealand*. Auckland: Environmental Defence Society.

Peart, R. and P. Reaburn. 2011. *Strengthening Second Generation Regional Policy Statements*. Auckland: Environmental Defence Society.

Severinsen, G. 2019. *Reform of the Resource Management System: A Model for the Future, Synthesis Report*. Auckland: Environmental Defence Society.

Severinsen, G. and R. Peart. 2018. *Reform of the Resource Management System: The Next Generation, Synthesis Report*. Auckland: Environmental Defence Society.

Chapter 16 Inside EDS

References

Jesson, B. 1991. Gary Taylor: The man who rose without trace. *Metro*, April: 112–9.

Riddell, M. 1987. Tinker, Taylor, *New Zealand Outlook*, March: 50–5.

Young, C. 1987. Svengali of the west; Gary Taylor. *Metro*, April: 98–105.

Glossary of Māori words

hapū sub-tribe

iwi tribe

kai food

kai moana food from the sea

kāinga home

kaitiaki guardian

mana whenua authority over land or territory

marae the open area in front of a meeting house for formal discourse, but also the complex of buildings around the marae

mātauranga Māori Māori knowledge

rangatiratanga the right to exercise authority

rohe region

rongōa medicine

rūnanga Māori council or assembly

taiāpure coastal reserve

tangata whenua people of the land

taonga treasures

te ao Māori the Māori world

te mana o te wai recognises spiritual power of water and the relationship between water, the wider environment and communities

tikanga customary system of values and practices

wāhi tapu sacred place

wāhi taonga sites of significance to Māori

whānau family

whenua land

Index